CORE HIGHER GRADE

# GEOGRAPHY
## QUESTIONS AND SKILLS

JOHN GREENLEES

Hodder & Stoughton

A MEMBER OF THE HODDER HEADLINE GROUP

## ACKNOWLEDGEMENTS

The publishers would like to thank the following for permission to reproduce materials in this book. Every effort has been made to trace and acknowledge all copyright holders, but if any have been overlooked, the publishers will be pleased to make the necessary arrangements.

Ordnance Survey mapping, with the permission of The Controller of Her Majesty's Stationery Office © Crown copyright, Map 4

Inside artwork by 1–11 Line Art

All photographs belong to the author.

*British Library Cataloguing in Publication Data*
A catalogue record for this title is available from The British Library

ISBN 0 340 69769 5

First published 1997

| Impression number | 10 | 9 | 8 | 7 | 6 | 5 | 4 | 3 | 2 | 1 |
|---|---|---|---|---|---|---|---|---|---|---|
| Year | 2002 | 2001 | 2000 | 1999 | 1998 | 1997 | | | | |



Typeset by Wearset, Boldon, Tyne and Wear.
Printed in Great Britain for Hodder & Stoughton Education, a division of Hodder Headline Plc, 338 Euston Road, London NW1 3BH by Scotprint Ltd, Musselburgh, Scotland.

# Contents

# INTRODUCTION

*Core Higher Grade Geography: Questions and Skills* provides a series of exam-style questions to help develop knowledge and understanding of Core Geography – the first and compulsory part of the Higher Grade Geography course.

The questions also provide invaluable practice at analysing and interpreting the maps, diagrams, graphs, tables of statistics, sketches, photographs and other information sources which are used in exam questions. A section on Information Handling Skills has been included to provide notes and advice on how to extract information from these sources and how to evaluate the information which is extracted. There are also notes and questions to help develop key Geographical Methods and Techniques, including collecting, analysing and presenting information.

For each of Core Geography's eight sub-sections there is a list of key topics and skills along with a detailed glossary to explain the meaning of key words. A set of model answers has been included at the back of the book to provide an indication of the standard of responses which are required for top grades.

## INFORMATION SOURCES

A wide range of information sources are used in the Higher Grade Geography course. These include:

- maps such as choropleth maps, isoline maps and Ordnance Survey extracts;
- graphs such as line graphs, bar graphs, triangular graphs and pie charts;
- diagrams such as cross-sections, transects and flow diagrams;
- sketches including field-sketches and models;
- photographs such as aerial photographs.

The following notes will help you identify these different information sources and show you how to extract information from them.

## MAPS

A choropleth map uses different types of shading to show variations in rainfall, population density and a variety of other measurements. The lighter shadings are usually used to show lower densities and numbers while the darker shadings usually show higher densities and numbers. Map 1 is a choropleth map which shows variations in annual rainfall totals in west Africa.

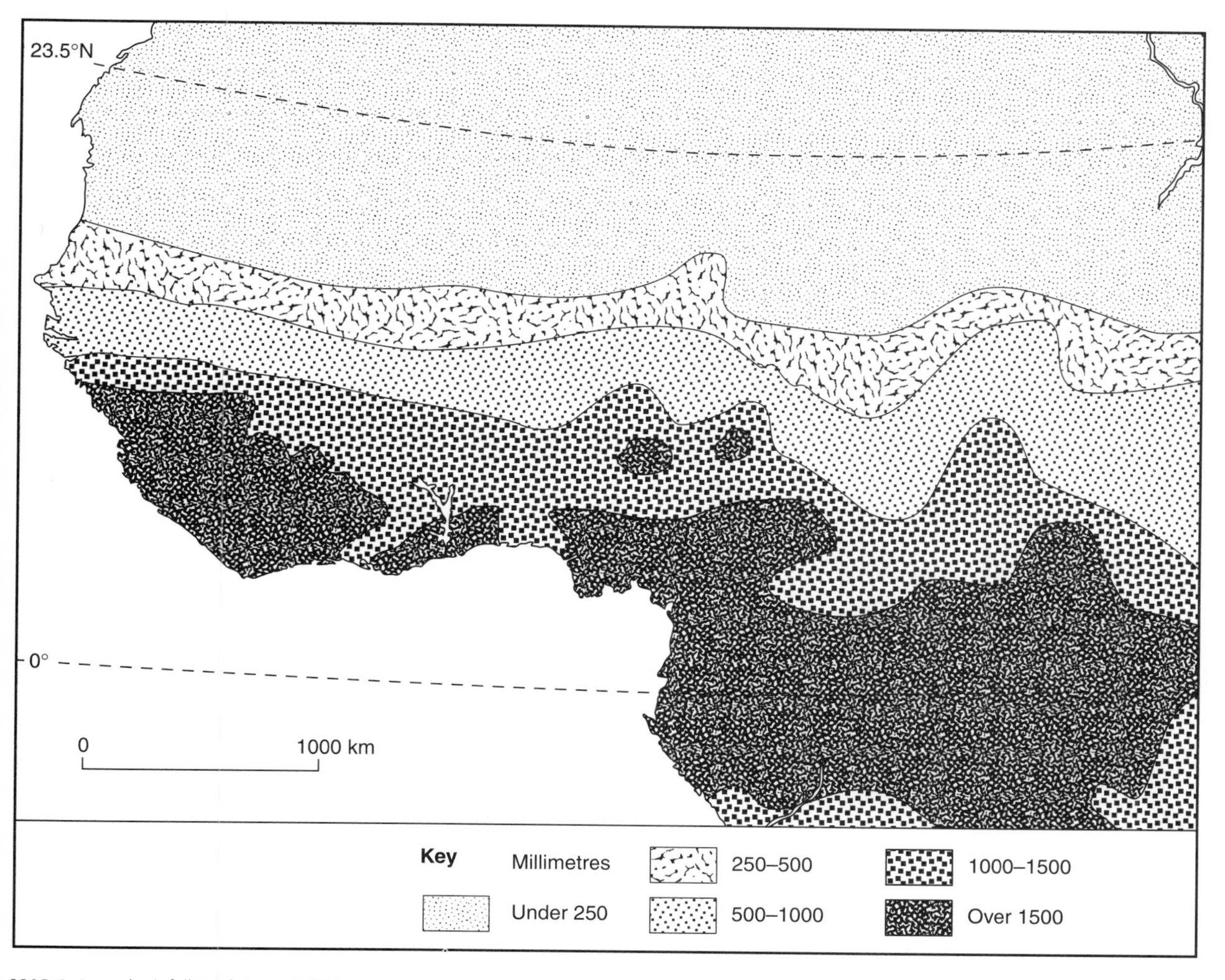


**MAP 1** Annual rainfall totals in west Africa

An isoline map uses lines to show places which have the same measurements, for example rainfall or temperature. Lines close together indicate greater change while lines far apart indicate more gradual change. A contour line, which joins places lying at the same height, is a good example of an isoline. Map 2 is an isoline map which shows variations in the accessibility of Stirling.

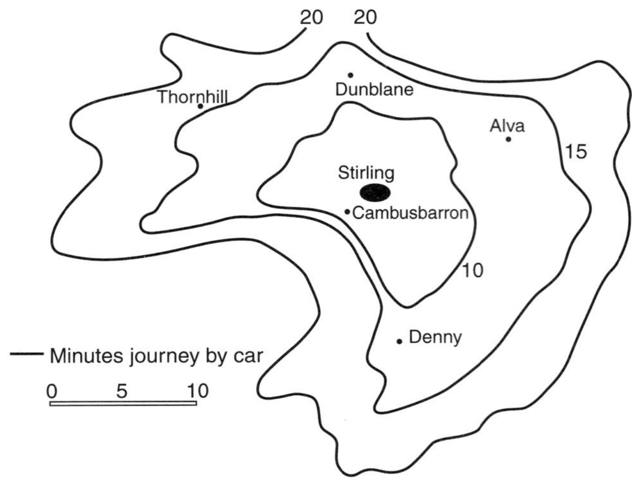


**MAP 2** The accessibility of Stirling

A sketch map is used to show the location of a particular feature or group of features, for example the location of factories, farms and rivers. Map 3 is a sketch map of the basin of the River Clyde.

An Ordnance Survey map shows a wide variety of physical and human landscape features. Physical features include landforms, drainage features and vegetation. The contour lines enable the relief (the shape and height) of the land to be studied and for cross-sections to be drawn. Human features such as settlements, factories and transport links are usually influenced by the relief of the land – transport links, for example, will avoid landforms such as steep slopes and mountains and make use of flat and low-lying areas such as plains and valleys.

On the Ordnance Survey map on the inside back cover of this book you should be able to:

- identify the following landforms: mountains, cliffs, valleys, corries, ridges, scree and a U-shaped valley;
- identify the following drainage features: rivers, tributaries, confluences, waterfalls, coastal flat/delta, watersheds and lochs;
- describe the site and situation of settlements and industrial facilities;
- identify and explain land uses such as roads, footpaths, forests and tourist facilities.

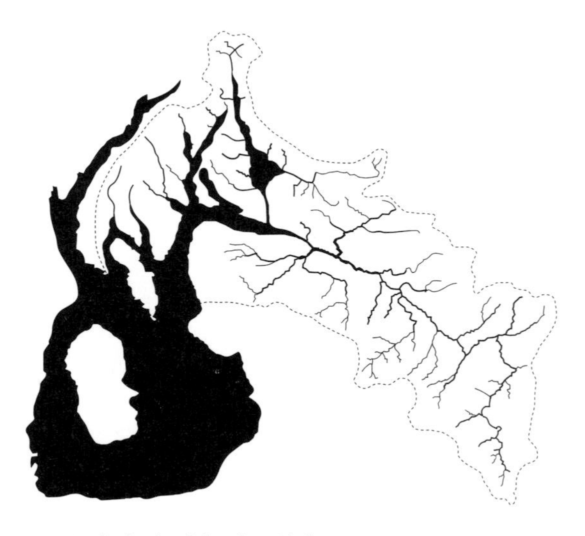

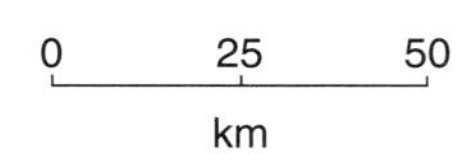


**MAP 3** The basin of the River Clyde

## GRAPHS

A line graph uses a single line to show changes over a period of time, for example Graph 1 is a line graph which shows the daily changes in the volume of road traffic in the centre of Aberdeen. Group line graphs include more than one line to enable comparisons to be made, for example Graph 2 is a group line graph which compares the birth rates of three countries.

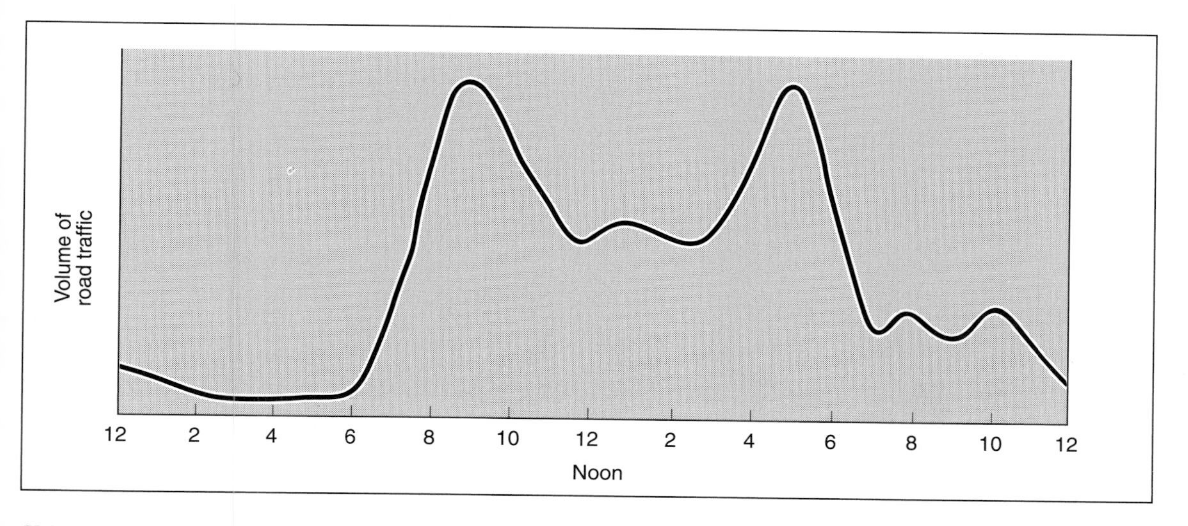


**GRAPH 1** Daily changes in the volume of road traffic in the centre of Aberdeen

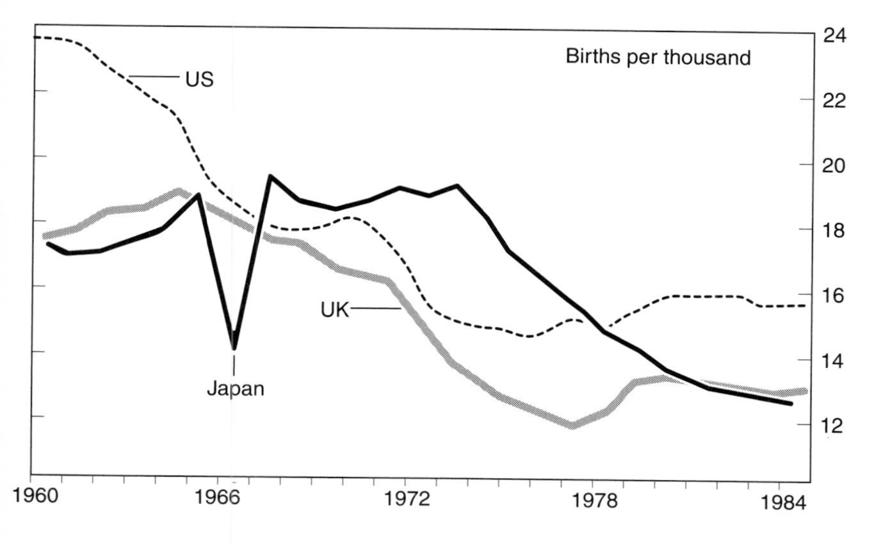


**GRAPH 2** Birth rates from three countries

Compound line graphs show the different component parts of a total, for example Graph 3 is a compound line graph which shows population growth in the developed world, the developing world and the world as a whole. Divergence line graphs show how measurements vary above and below an average or some other value, for example Graph 4 is a divergence line graph which shows global temperature change.

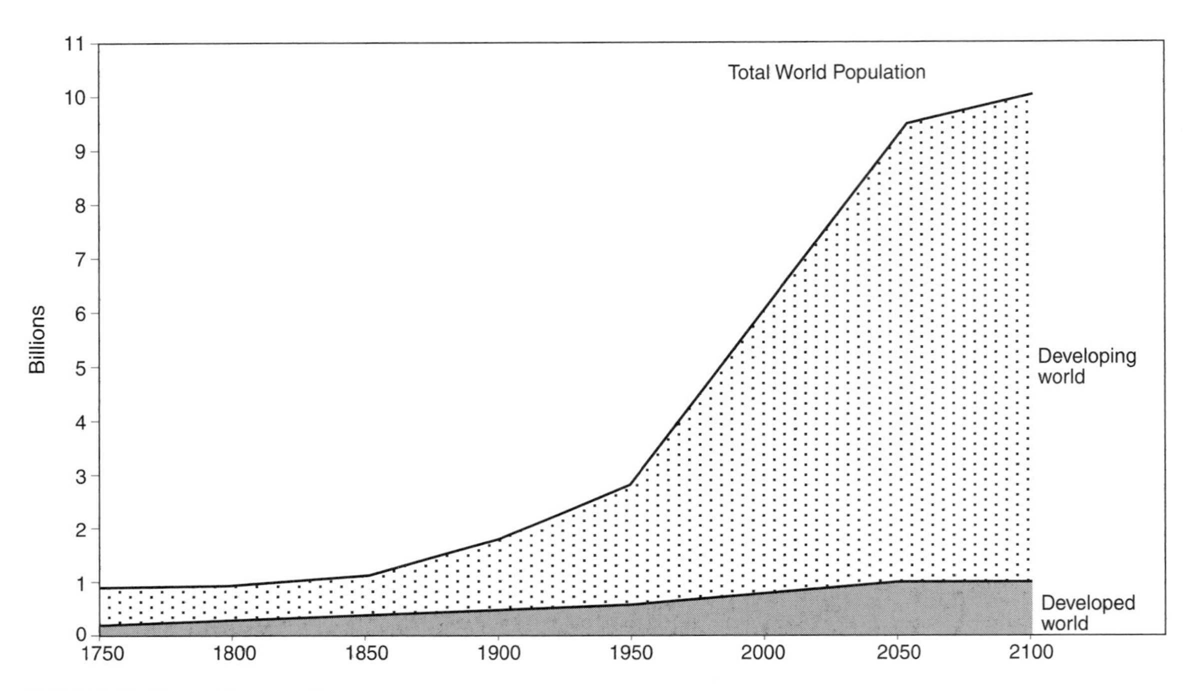


**GRAPH 3** World population growth

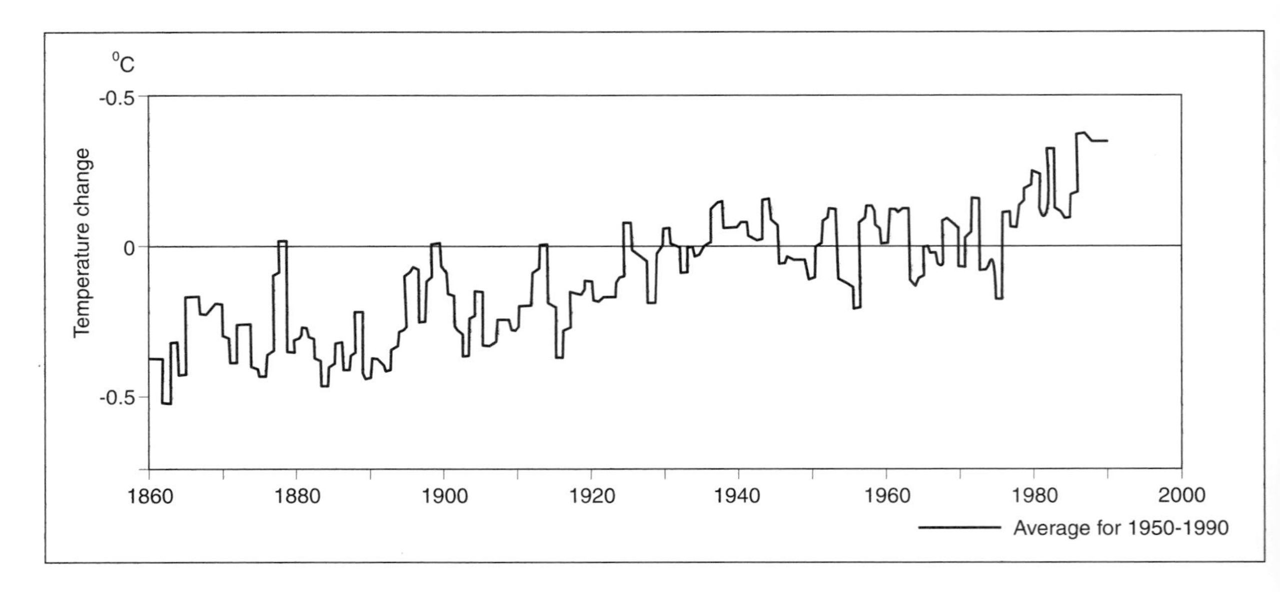


**GRAPH 4** Global temperature change

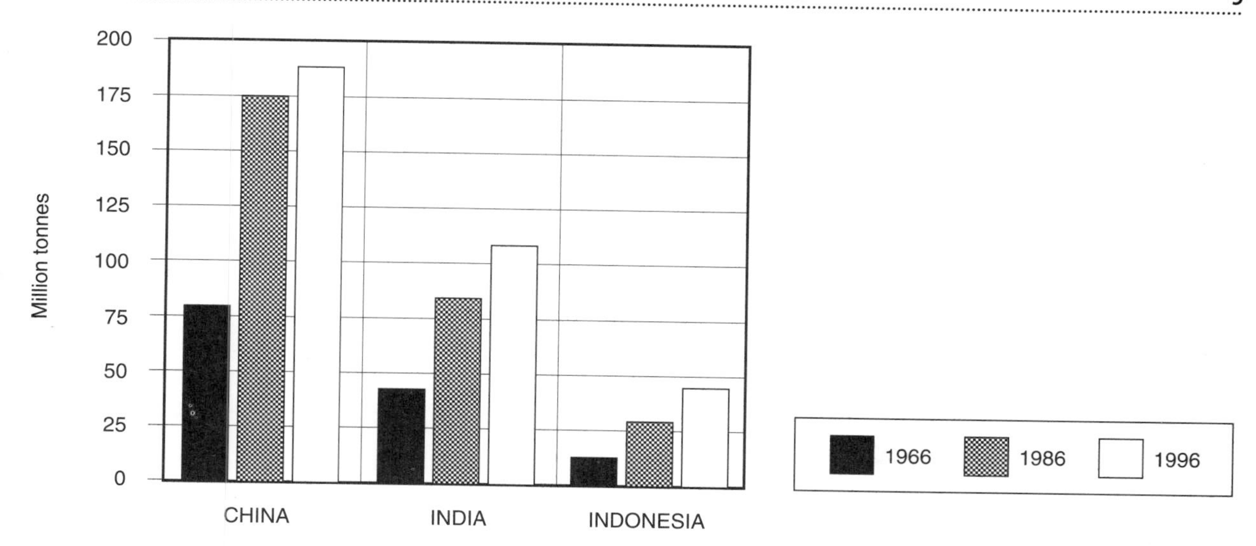


**GRAPH 5** Changes in rice production in three countries

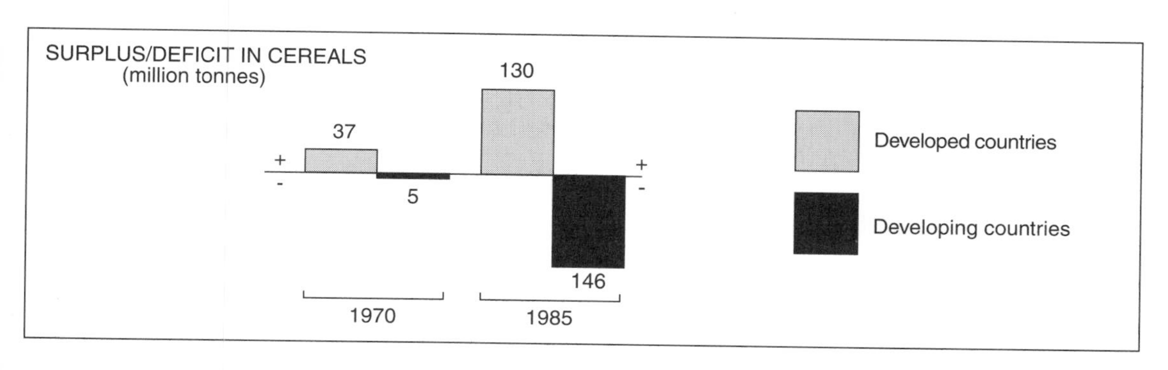


**GRAPH 6** Surpluses and deficits in cereals

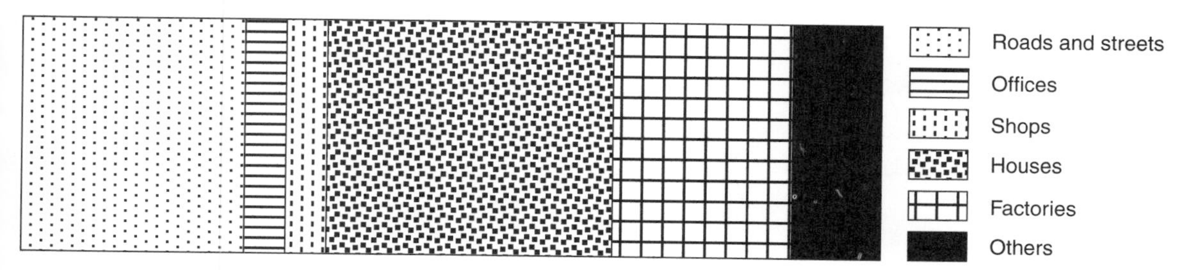


**GRAPH 7** Land use in Glasgow's inner city

A bar graph uses a number of bars to compare different measurements, for example Graph 5 shows changes in rice production in three countries. Divergence bar graphs show how measurements vary above and below an average or some other value, for example Graph 6 is a diveregnce bar graph which shows surpluses and deficits in cereals in developed countries and developing countries.

A divided bar graph shows the different parts of a total value, for example Graph 7 is a divided bar graph which shows land use in Glasgow's inner city.

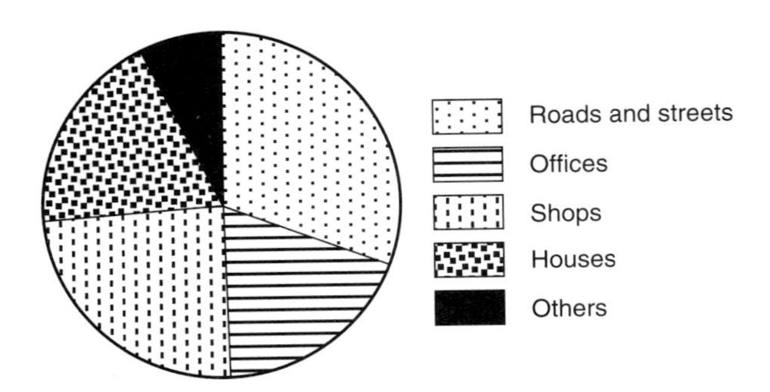


**GRAPH 8** Land use in Glasgow's CBD

A pie chart is a circular form of a divided bar graph, for example Graph 8 is a pie chart which shows land use in Glasgow's central business district.

A scatter graph shows the relationship between two variables, for example Graph 9 is a scatter graph which shows the relationship between life expectancy and natural population increase.

A triangular graph has three different axes – an X axis, a Y axis and a Z axis – to show three different sets of data. A dot or cross is marked at the point where the three values meet. To read a value you follow the nearest line to the edge of the triangle. The three sets of data must add up to the same values (100 per cent is the most common). Triangular graphs, such as Graph 10, are often used to compare types of employment in different countries.

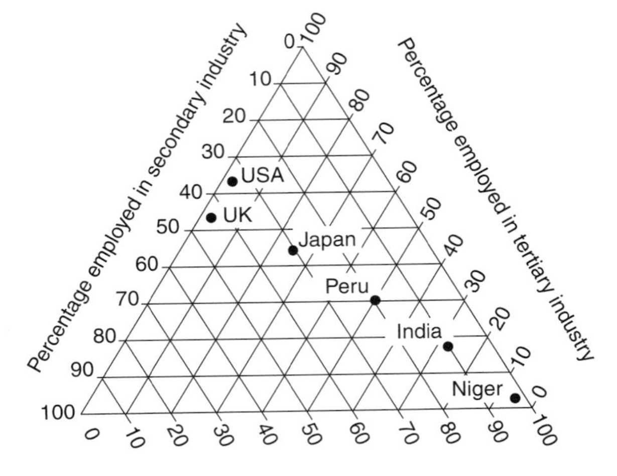


**GRAPH 10** Employment in six countries

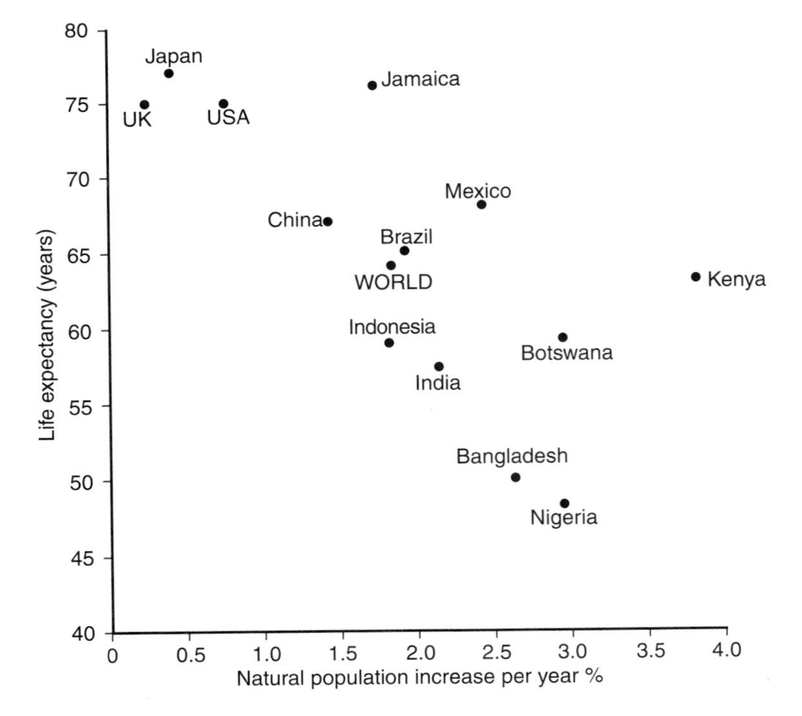


**GRAPH 9** The relationship between life expectancy and natural population increase

# DIAGRAMS

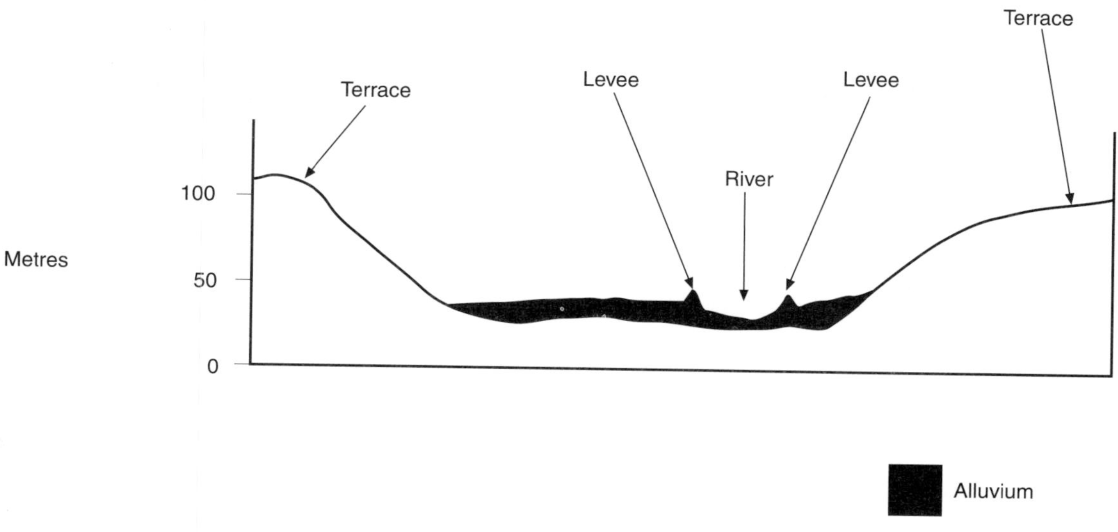


**DIAGRAM 1** A flood plain

A cross-section (Diagram 1) provides a sideways view of an area of land to enable the shape and height to be studied. A transect (Diagram 2) is a cross-section with additional information about geology, landforms, land use, etc.

A flow diagram (Diagram 3) shows the movements and values of movements between places. The direction of movement is shown by an arrowhead while the number or amount involved is shown by the width of the line. Flow diagrams are often used to show transport flows and population migrations, for example Diagram 3 shows traffic flows in the Stirling area.

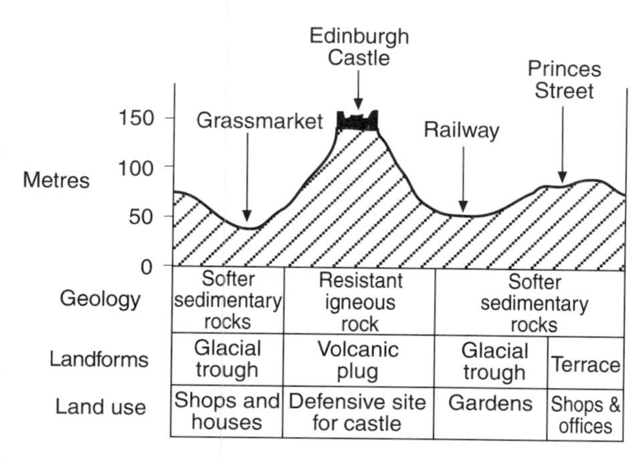

**DIAGRAM 2** Central Edinburgh

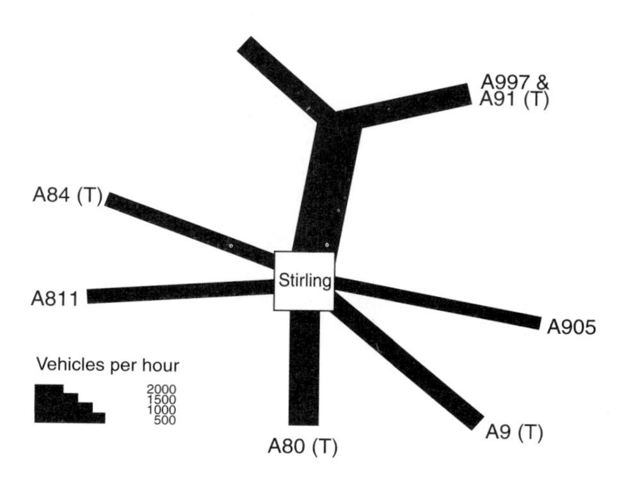


**DIAGRAM 3** Traffic flows in the Stirling area

## SKETCHES, MODELS AND PHOTOGRAPHS

Sketches enable key landscape features to be identified and highlighted. A field sketch is a sketch which is drawn from observations made during a field trip. An annotated sketch is a sketch which has labels and notes to provide additional information about key features. Field sketches should contain a title, a north point and an indication of scale. Sketch 1 shows a field sketch of Loch Leven (the area partly covered by the OS map on the inside back cover).

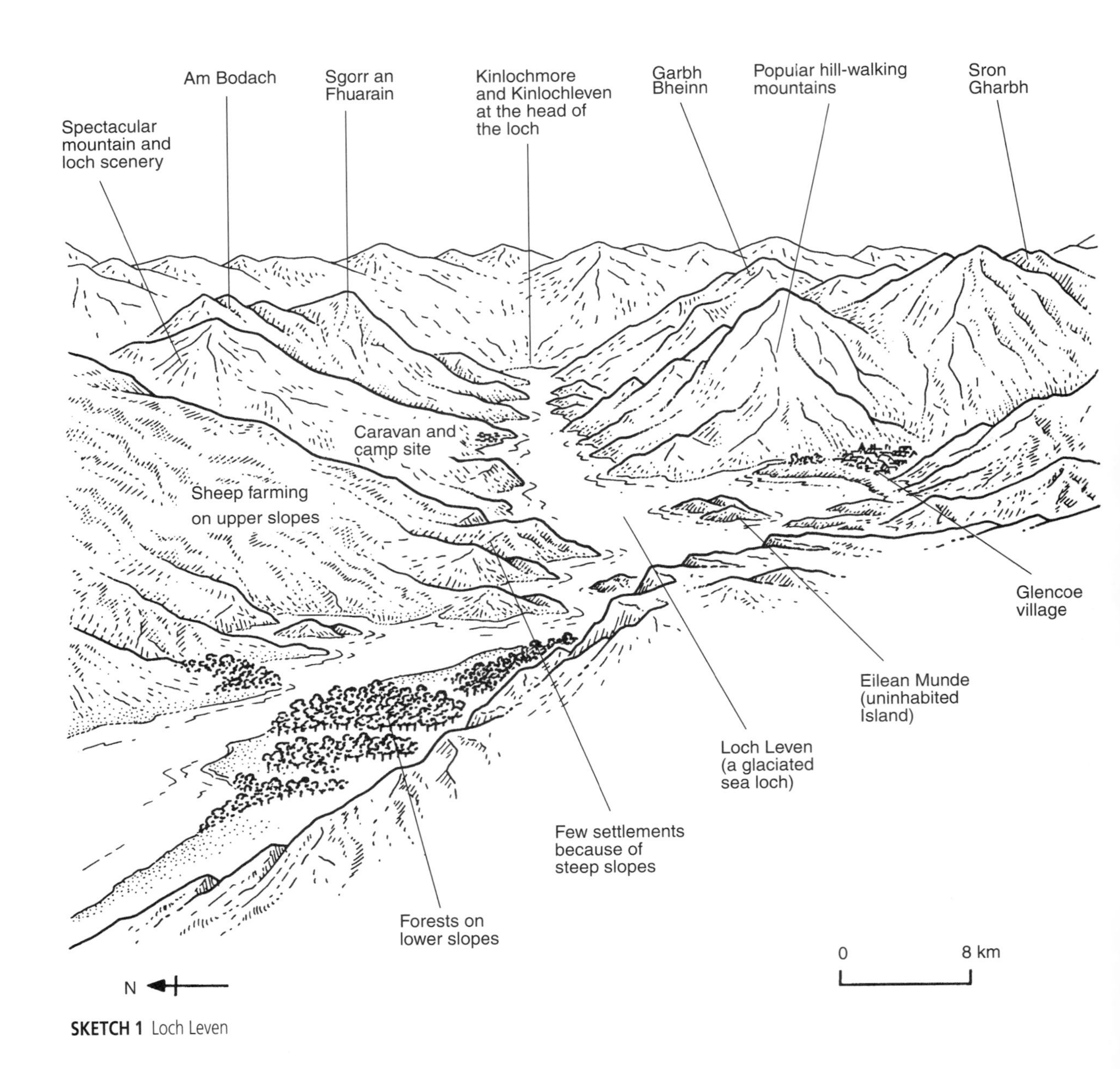


**SKETCH 1** Loch Leven

A model is a sketch which shows what could be expected or what might happen. Model 1 shows a model of land use in a city.

An aerial photograph shows a view from the air to enable physical and human landscape features to be studied. Photograph 1 shows an aerial photograph of an urban landscape.

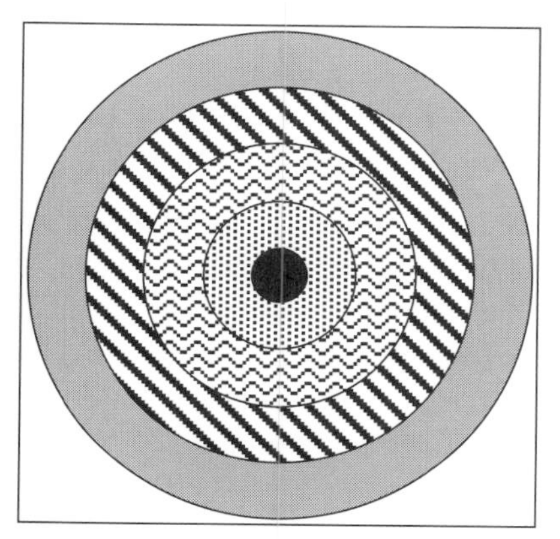

**MODEL 1** Model of land use in a city

## SOME GENERAL TIPS FOR EXTRACTING INFORMATION FROM INFORMATION SOURCES

- Look at the title of the information source to find out more about what it shows.
- Look at the key, if there is one, to establish what the different shades and symbols mean.
- Read the information along the sides of graphs to identify the units of measurement involved – remember to use these units of measurement if you include any data from the graph in your answer.
- Identify the key patterns or trends shown by the information source, for example 'a steep rise', 'a moderate fall', 'fluctuating movements' and 'no change'.

**PHOTOGRAPH 1** An urban landscape

Geographical methods and techniques are an important part of Core Geography. These methods and techniques include:

- constructing and analysing graphs;
- constructing and interpreting cross-sections and transects;
- analysing soil profiles, land use data and the results of surveys;
- interpreting data;
- annotating and analysing field sketches.

## CONSTRUCTING AND ANALYSING A CLIMATE GRAPH

A climate graph shows the average monthly temperature and precipitation figures for one weather station. Climate Graph 1 shows the temperature and precipitation figures for the weather station at Edinburgh. Figures are collected over a long period – at least 40 years – to give an accurate representation of the area's climate. An analysis of a climate graph should provide a note of the key features of the climate, including:

- the warmest and coldest months;
- the highest and lowest temperature figures;
- the range of temperature;
- the wettest and driest months;
- the highest and lowest monthly precipitation figures;
- the total annual precipitation figure;
- the months which make up the hot season, the cold season, the wet season and the dry season.

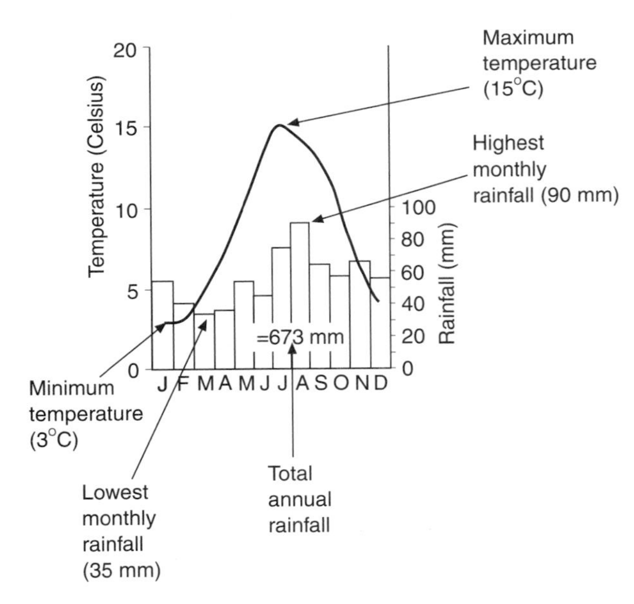


**CLIMATE GRAPH 1** The climate of Edinburgh

The range of temperature (the difference between the highest and lowest monthly temperatures) can be calculated by subtracting the smallest monthly temperature figure from the highest monthly temperature figure. The total annual precipitation figure can be calculated by adding up the 12 monthly precipitation figures.

## CONSTRUCTING AND ANALYSING A HYDROGRAPH

A hydrograph shows changes in the discharge of a river. The discharge of a river is the quantity of water which flows past a fixed point during a certain time. This is usually measured in cubic metres per second or, in some cases, by the changing depth of the river. The base flow is the normal flow of water while the storm flow is the additional water which results after heavy rain. The pattern of seasonal changes in the discharge of the river is called the regime.

A flood hydrograph (Hydrograph 1) shows the effect of heavy rainfall on the river's discharge. The hydrograph will show a lag time – the time gap between the peak rainfall and the peak discharge. The length of the lag time will depend on the characteristics of the rock, soil and vegetation in the river's basin and how much water they are holding at the time of the rainfall. One of the main characteristics of a flood hydrograph is the steep rise to peak discharge followed by a slower fall as the flood waters subside. The steep rise is known as the rising limb while the slower fall is known as the recession limb.

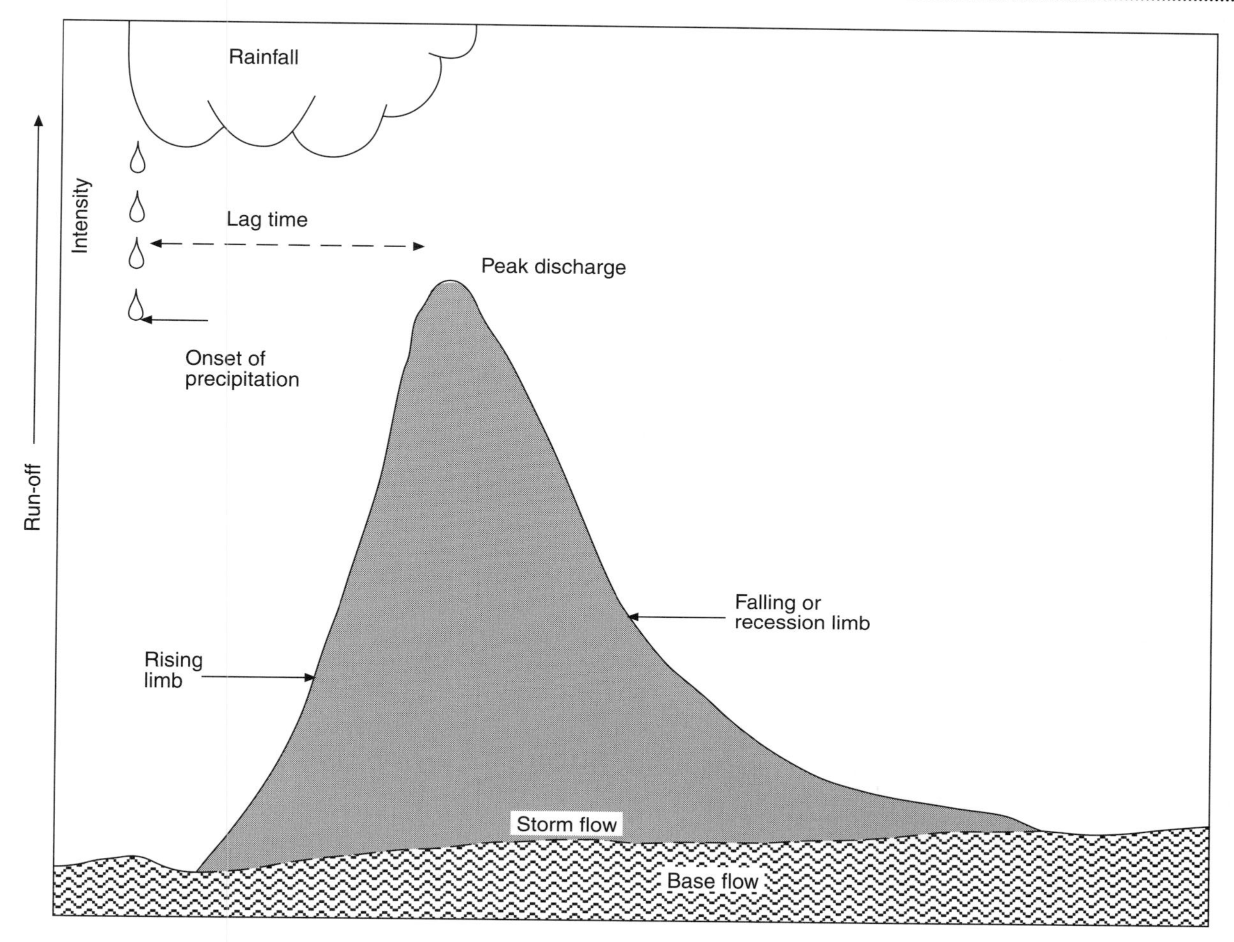


HYDROGRAPH 1

The steepness of the rising limb will be dependent on the speed of the transfer of water to the river. Human activities, such as planting and removing trees, can affect the steepness of the rising limb. Planting trees will reduce the amount of water running off the land and reduce the steepness of the rising limb. Removing trees will have the opposite effect and increase the chances of flooding. The steepness of the recession limb, which represents the water running out of the river system, is determined by the nature of the river basin.

Water engineers can use hydrographs to predict when flooding is likely to occur, for example when the lag time is particularly short and the peak is particularly high.

## CONSTRUCTING AND INTERPRETING A CROSS-SECTION AND TRANSECT

A cross-section between two points can be constructed as follows.

- Measure the distance between the two points – let us call them points A and B.
- Use the edge of a piece of paper to mark in the position, and the heights, of the contour lines which lie between points A and B (for very steep areas only mark in every fifth contour line).
- Draw a line the same length as the distance between A and B on a piece of graph paper.
- Add a vertical scale from 0 metres to a height

which is at least the same height as the highest land on your section.

- Plot the points you marked along the edge of the piece of paper on to your cross-section frame. Join up the points to produce a cross-section.
- You can then add a title, a note of the direction and the names of important features such as rivers or mountains.

The completed cross-section (Cross-section 1) enables the shape and height of the land to be studied and for features such as U-shaped valleys and flood plains to be identified.

Additional information obtained from field or library investigations can be added below the cross-section to form a transect. Transects enable links and relationships to be established, for example the influence of structure and rock type on soils, relief and land use. Low relief and fertile soils will probably be linked to land uses such as arable farming, while high relief and poorer soils are likely to be linked to land uses such as sheep farming and forestry.

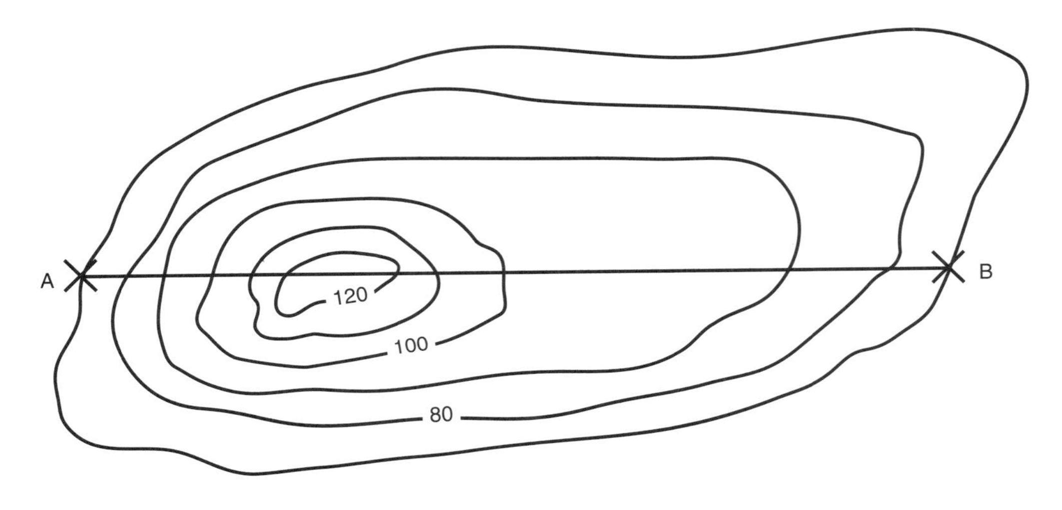


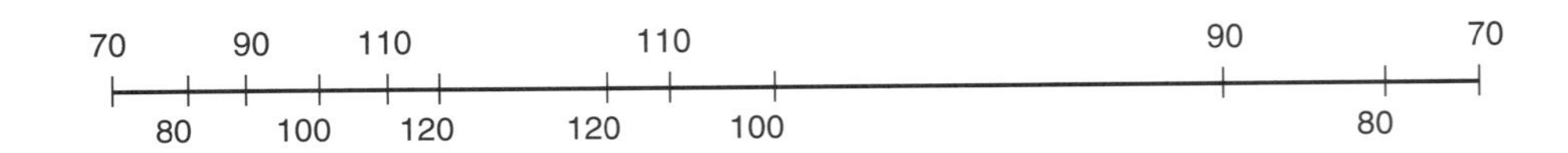

Drawing a cross-section

## ANALYSING A SOIL PROFILE

A soil profile is a type of divided bar diagram which shows the structure of a soil. The different bands which are shown in the soil profile are called horizons – the A-horizon, the B-horizon and the C-horizon. Analysing a soil profile will enable you to identify the properties of a soil and help you to understand some of the processes which operate within the soil.

In Soil Profile 1 there is a thick ground cover of leaves which have fallen from deciduous trees. These leaves provide the A-horizon with sufficient humus (decayed organic matter) to make the soil suitable for growing a wide range of crops including wheat and barley. There is a net downward movement of moisture and minerals through the soil so that some leaching (the removal of dissolved chemicals) occurs in the B-horizon. The iron pan, which has started to develop at the top of the B-horizon, might eventually impede drainage and cause some waterlogging in the A-horizon. The roots of trees, however, are able to penetrate well down into the soil to absorb moisture and nutrients. Soil Profile 1 is a brown forest soil.

Soil Profile 2 shows a soil with a thin layer of raw acid humus deposited by coniferous trees. This makes it a largely infertile soil with limited value for growing crops. The net downward movement of moisture causes heavy leaching which leaves behind an orange-brown iron pan at the top of the B-horizon. This iron pan can impede drainage and cause waterlogging in the A-horizon. Soil Profile 2 is a podzol.

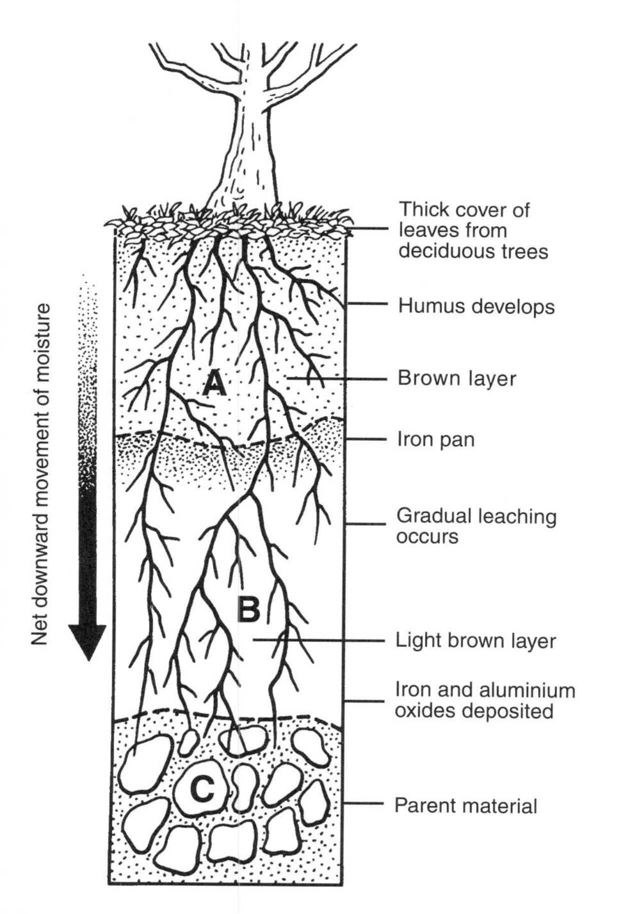


SOIL PROFILE 1

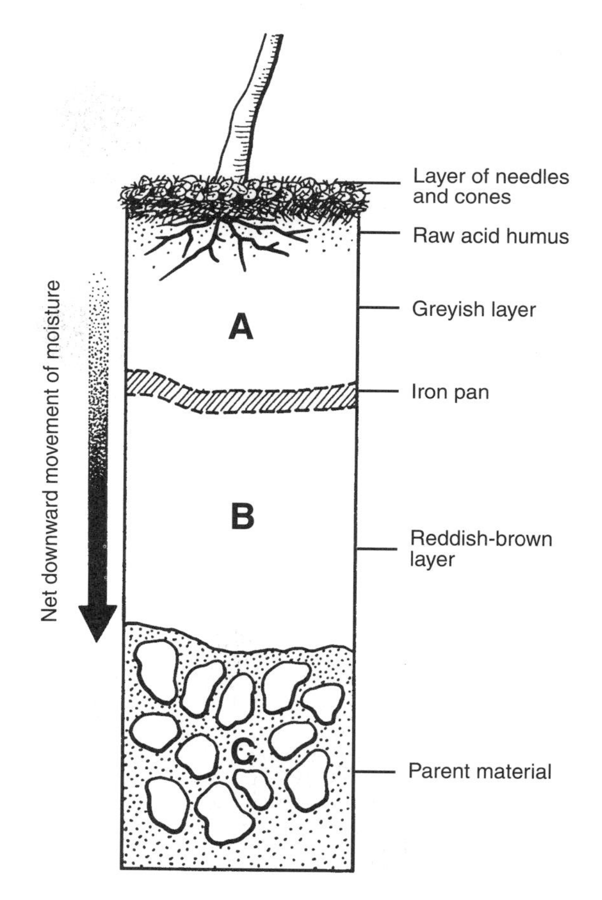


SOIL PROFILE 2

## INTERPRETING POPULATION DATA

A population pyramid is a form of bar graph which shows the age and sex structure of a population. Ages are usually based on five-year periods (nought to four years, five to nine years, etc.), with the youngest age groups at the base of the graph. Males are shown on one side and females on the other.

The shape of a population pyramid will help you identify whether the country is a developing country or a developed country. Developing countries have broader bases and narrower tops because of their high birth rates, high death rates and lower life expectancy figures. Developed countries will have narrower bases and broader tops because of their low birth rates, low death rates and longer life expectancy figures.

Variations of the model population structure can be explained by a number of different possibilities. Fewer people in the middle groups may be the result of emigration, while more people in these groups may be the result of immigration. Fewer people in certain groups may be the result of a war or a natural disaster such as a flood or famine.

The population pyramid therefore enables problems, such as a high birth rate or a high death rate, to be identified. Population Pyramids 1 to 5 show five different population patterns:

- an expanding population;
- a contracting population;
- a stable population;
- a stable population after major war losses;
- an expanding population with immigration of young adult males.

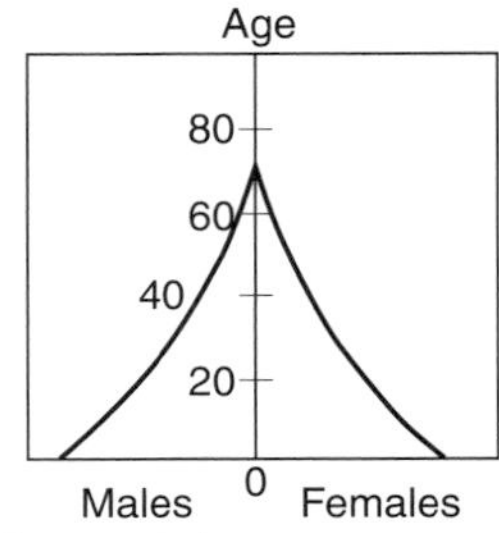

**POPULATION PYRAMID 1** An expanding population

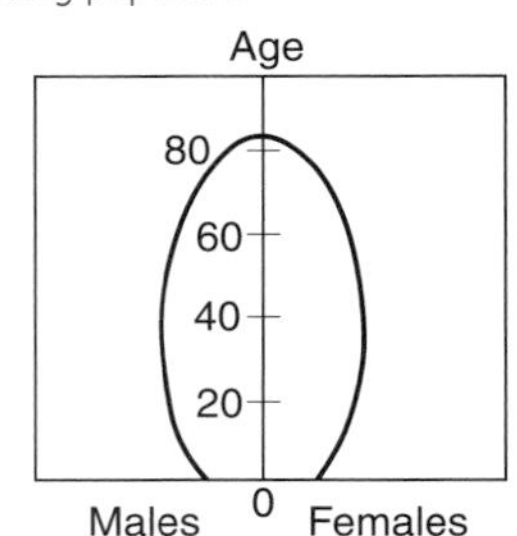

**POPULATION PYRAMID 2** A contracting population

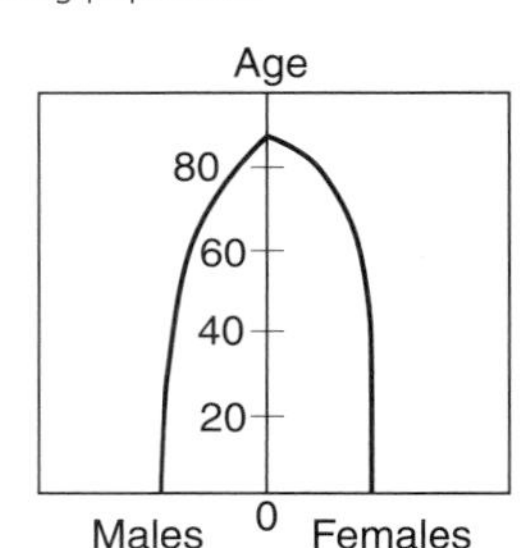

**POPULATION PYRAMID 3** A stable population

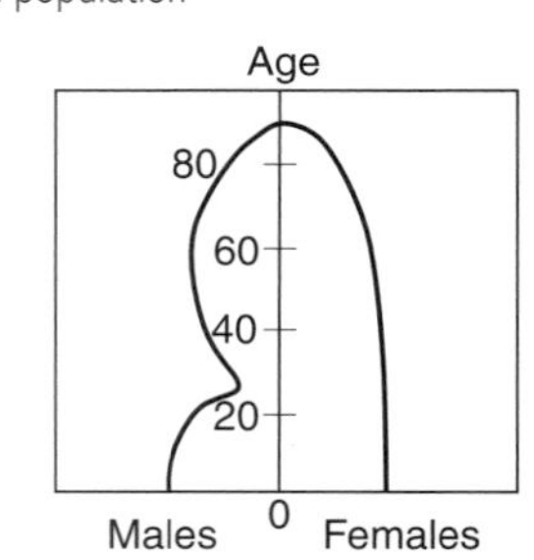

**POPULATION PYRAMID 4** A stable population after major war losses

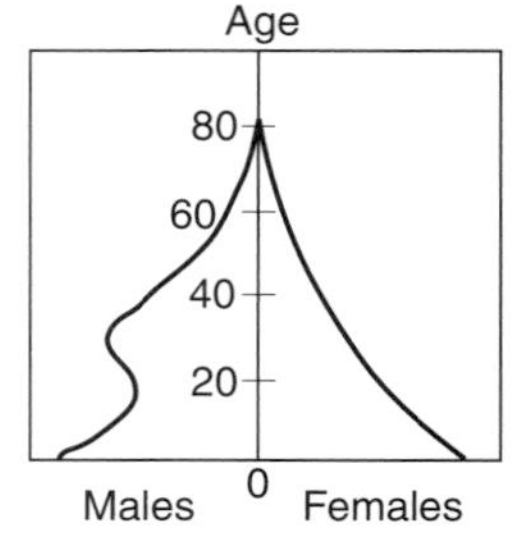

**POPULATION PYRAMID 5** An expanding population with immigration of young adult males

## ANALYSING LAND USE DATA

Land use is the way in which the land is being used. In rural areas the main land uses are farming, forestry, quarrying, recreation, transport and small settlements. In urban areas the main land uses are large settlements, industry, retailing, services, transport and recreation. Land use surveys enable landscape uses and changes to be studied and analysed. The survey data below shows how land use in an area in East Anglia changed between 1945 and 1995.

| | 1945 % | 1995 % |
|---|---|---|
| wheat | 26 | 58 |
| barley | 24 | 0 |
| peas | 0 | 22 |
| sugar beet | 12 | 0 |
| other crops | 4 | 0 |
| woodland | 8 | 0 |
| hedgerows | 16 | 4 |
| settlements | 6 | 9 |
| roads | 4 | 7 |

An analysis of these land use data shows that the area concerned is a rural area in which arable farming is the dominant economic activity. In 1995 two cash crops – wheat and peas – accounted for 80 per cent of the area's land use while settlements and roads accounted for only 16 per cent of the land use.

The data also show that the area has experienced major changes in land use. Between 1945 and 1995 the amount of land taken up by wheat and peas increased from 26 per cent to 80 per cent while the amount of land used for growing barley, sugar beet and various other crops dropped from 40 per cent to 0 per cent. Trees have also disappeared from the area and the number of hedgerows has decreased significantly. The amount of land used for roads and settlements has increased slightly.

The changes in crops suggest that the area's farmers have switched from growing a broad range of crops to specialising in what are probably the area's most profitable crops – wheat and peas. The increase in pea production may also be explained by the opening of a new frozen pea factory nearby. The disappearance of trees and hedgerows is due to increased mechanisation on arable farms and the need for larger fields to accommodate large machines such as combine harvesters. The increase in the number of roads reflects the increase in traffic which has occurred and the need for new transport routes including bypasses and motorways. Settlements have expanded because of natural population growth and the demand for larger houses.

## ANALYSING THE RESULTS OF INDUSTRIAL SURVEYS

Industrial surveys provide useful information about the different industries which are important to an area. The results of a survey enable industrial changes to be studied over a period, for example the decline in coalmining, steelmaking and shipbuilding in central Scotland and the growth of electronics, computers and other high-tech industries.

Surveys will also indicate the different types of industries (primary, secondary and tertiary) which are important to a town, region or country and allow comparisons to be made with other towns, regions or countries. The data below, for example, show that primary industries such as farming provide the greatest number of jobs (75 per cent) in Country X while manufacturing and tertiary jobs are less important. This would suggest that Country X is a developing country.

The data also show that Country Y has a well-developed secondary or manufacturing sector with only a small number (8 per cent) of jobs in primary activities. This would suggest that Country Y is a developed country.

The survey data suggests that Country Z is also a developed country. Primary jobs only account for 5 per cent of the country's workforce and most workers are employed in the manufacturing and tertiary sectors. Service sector jobs account for the biggest category of jobs in Country Z. This suggests that Country Z has a post-industrial type of economy in which highly-mechanised and profitable industries are very important but don't employ as many people as older traditional industries. The highly-mechanised and profitable nature of Country Z's industries enables the country to support a large number of service sector jobs.

| | primary | secondary | tertiary |
|---|---|---|---|
| Country X | 75 | 15 | 10 |
| Country Y | 8 | 62 | 30 |
| Country Z | 5 | 39 | 56 |

## ANNOTATING AND ANALYSING URBAN FIELD SKETCHES

Field sketches provide a record of observations made during field investigations in urban areas. Sketches can be annotated (labels and notes added) to provide additional information about the landscape. Annotations should include a title, a north arrow and a scale. An old urban landscape will show old buildings and factories mixed in together, while a new urban landscape will show separate housing neighbourhoods and industrial estates. A changing urban landscape will include derelict land, open spaces and a combination of old and new buildings.

Field Sketch 1 shows an old highly-industrialised urban landscape which has developed next to a river. A riverside location is obviously important for industries such as shipbuilding and ship repairs as well as industries which require water for manufacturing purposes. The river also provides a means of transporting heavy industrial goods such as coal and steel. A number of workers' houses have been built within walking distance of the industrial sites. These houses date from an age when transport services were poor and workers had to walk to their workplaces. Space is limited and a large number of houses, without gardens or other green areas, have been built in a small area. Overcrowding is a problem. The industrial activities have had a major impact on the urban environment. There is evidence of water, land and air pollution from industries such as coalmining and shipbuilding. The residents of the nearby houses are probably affected by dust and noise pollution.

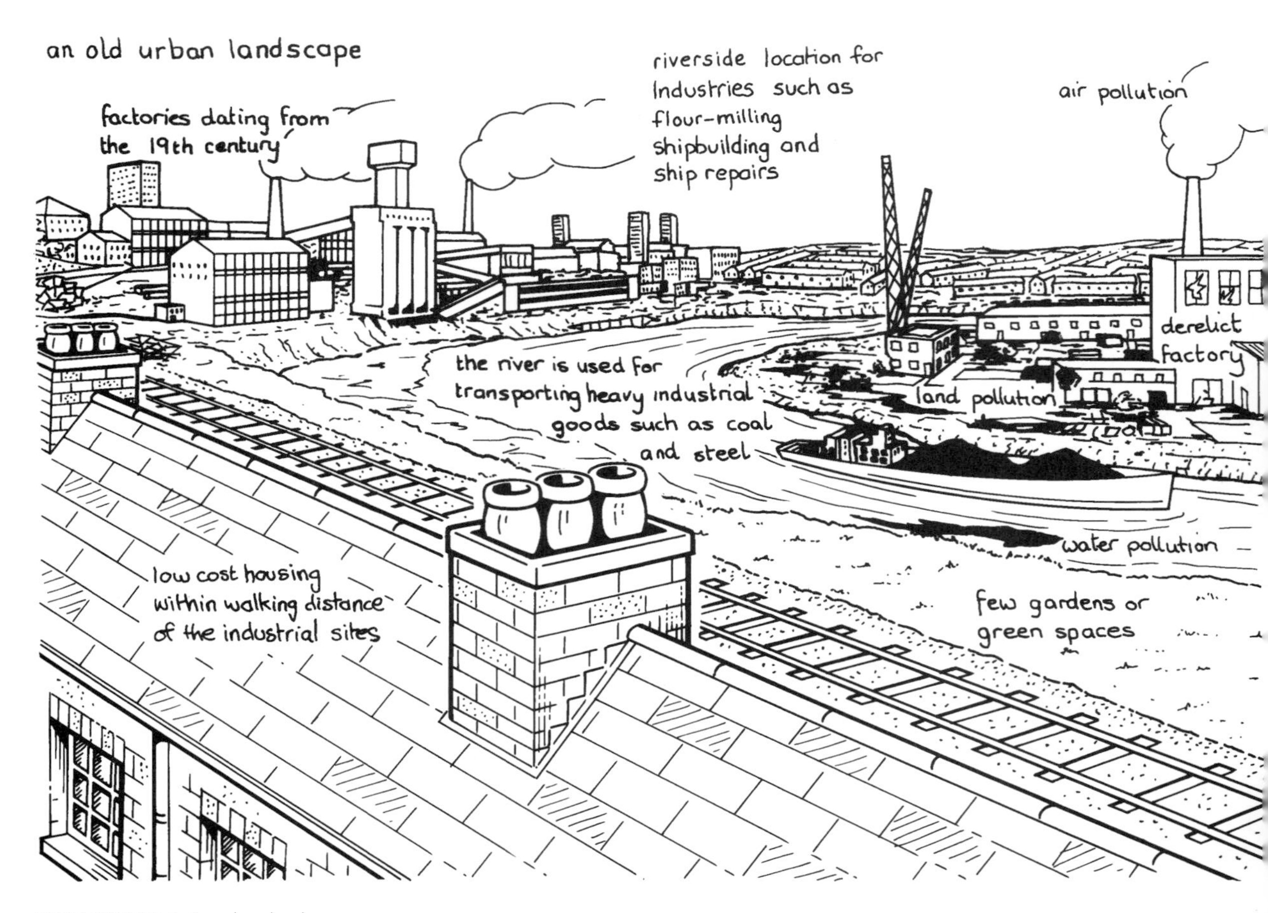


**FIELD SKETCH 1** An urban landscape

## QUESTIONS

**1** Figure 1.1 is a diagram showing the energy exchanges which occur at the earth's surface.

**(a)** Describe and explain the energy exchanges shown by the diagram. (6)

**(b)** Explain how human activities can affect the earth's energy exchanges. (4)

**(10)**

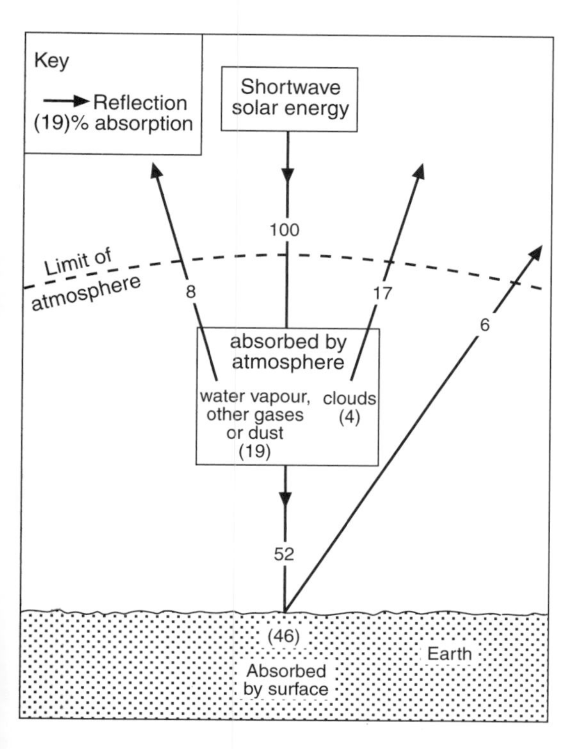


**FIGURE 1.1** The energy exchanges which occur at the Earth's surface

**2** Study Figure 1.2, which shows the earth's energy budget.

**(a)** Explain why low latitudes receive more insolation than high latitudes. (4)

**(b)** Describe how energy is transferred from low latitudes to high latitudes. (6)

**(10)**

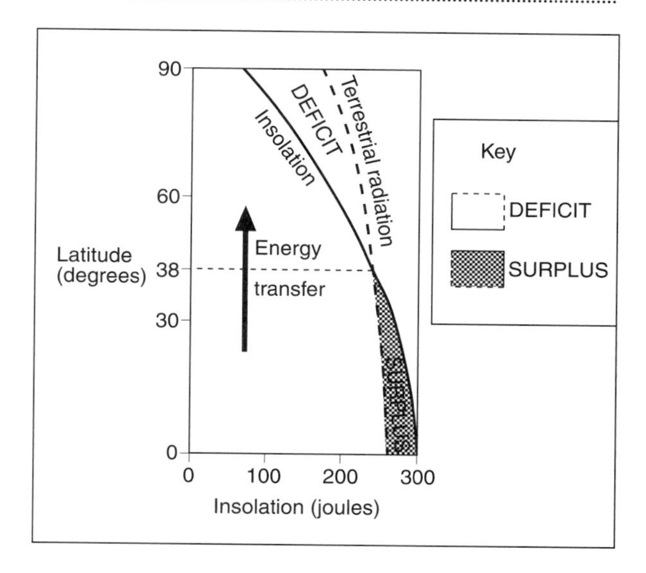


**FIGURE 1.2** The Earth's energy budget

**3** Look at Figure 1.3 which shows atmospheric circulation and surface winds.

**(a)** Describe the pattern of surface winds shown on the map. (5)

**(b)** Describe how atmospheric circulation helps to redistribute energy over the planet. (5)

**(10)**

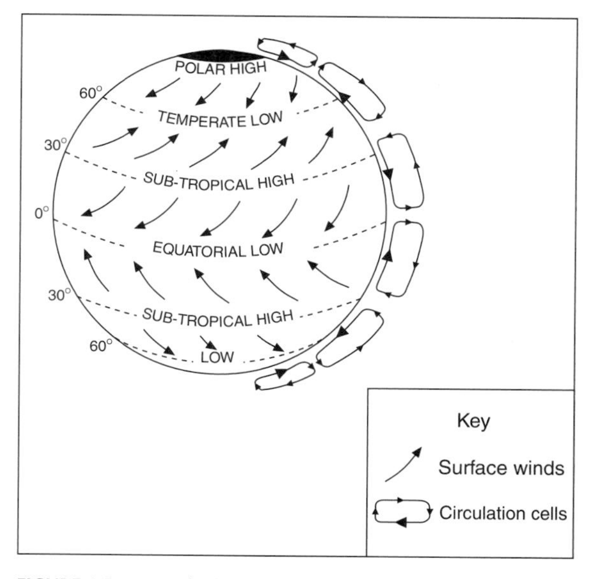


**FIGURE 1.3** Atmospheric circulation and surface winds

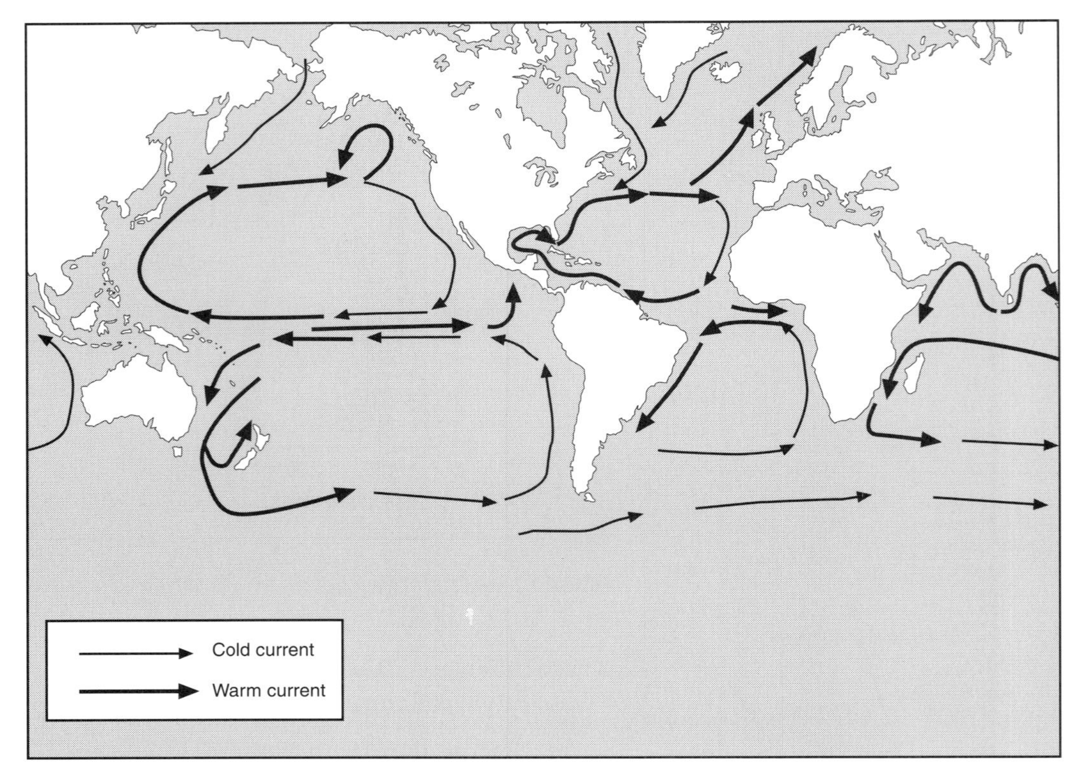


**FIGURE 1.4** Some of the World's ocean currents

**4** Look at Figure 1.4, which shows some of the world's ocean currents.

**(a)** Name one warm ocean current and one cold ocean current shown on the map. (2)

**(b)** Choose one of these ocean currents and describe its effects. (4)

**(c)** Outline the role played by all the world's ocean currents in redistributing energy over the globe. (4)

**(10)**

**5** Read Figure 1.5, which is part of a newspaper article about global climate change.

**(a)** Describe the human causes of global warming. (6)

**(b)** Outline the steps which can be taken to tackle the problem of global warming. (4)

**(10)**

## 'It's Getting Warmer'

A group of leading scientists has predicted that global temperatures will rise by as much as three degrees Celsius during the next century. In a report to the United Nations the scientists pointed out that the 1980s and 1990s provided some of the warmest years ever recorded. 'Higher average temperatures' they warned, 'will lead to melting in the polar ice caps and an accompanying rise in sea-levels.' 'Coastal areas will be inundated with seawater if no action is taken to slow or stop the planet's warming process.'

**FIGURE 1.5** A newspaper article

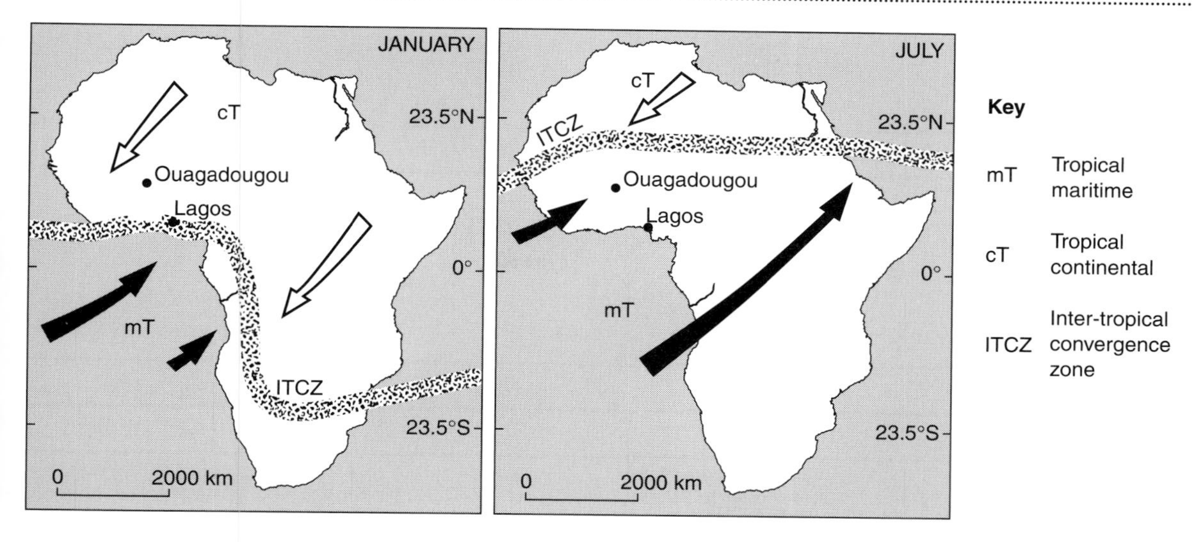


**FIGURE 1.6** The presence of air masses and fronts over Africa

**6** Study Figure 1.6, which shows the presence of air masses and fronts over Africa.

**(a)** Describe the characteristics of Tropical Maritime and Tropical Continental air masses. (5)

**(b)** Account for the characteristics of Tropical Maritime and Tropical Continental air masses. (5)

**(10)**

**7** Look at Figure 1.7 which shows rainfall in West Africa.

**(a)** Describe the pattern of rainfall in West Africa. (5)

**(b)** Explain how the Inter-Tropical Convergence Zone affects the distribution of rainfall in West Africa. (5)

**(10)**

**FIGURE 1.7** Rainfall in West Africa

Station X

| | J | F | M | A | M | J | J | A | S | O | N | D |
|---|---|---|---|---|---|---|---|---|---|---|---|---|
| temperature (°C) | 30 | 30 | 30 | 30 | 29 | 29 | 28 | 28 | 28 | 28 | 29 | 29 |
| rainfall (mm) | 100 | 120 | 40 | 145 | 170 | 110 | 135 | 220 | 190 | 240 | 160 | 80 |

Station Y

| | J | F | M | A | M | J | J | A | S | O | N | D |
|---|---|---|---|---|---|---|---|---|---|---|---|---|
| temperature (°C) | 33 | 33 | 33 | 33 | 32 | 31 | 30 | 31 | 35 | 36 | 36 | 34 |
| rainfall (mm) | 0 | 0 | 0 | 0 | 12 | 31 | 52 | 42 | 18 | 0 | 0 | 0 |

**FIGURE 1.8** Temperature and rainfall figures for two weather stations in West Africa

8 Figure 1.8 shows temperature and rainfall figures for two weather stations in West Africa.

**(a)** Construct two climate graphs to show a comparison of these figures. (4)

**(b)** Describe and account for the contrasts in rainfall shown by the graphs. (6)

**(10)**

TOTAL NUMBER OF MARKS **80**

# 2 Hydrosphere

## Key topics

- the components of the global hydrological cycle;
- areas of water sufficiency and deficiency in the Savanna regions of Africa north of the Equator;
- the effects of climate variability in the Savanna regions of Africa north of the Equator;
- the movement of water within drainage basins – inputs, storage and outputs;
- the effects of flowing water, including erosion, transportation and deposition;
- the upper, middle and lower courses of rivers;
- river landforms.

## Key skills

- identifying river features on OS maps;
- presenting river flow data;
- constructing and analysing hydrographs.

## Key words

**alluvial fan** a flat area of sediment which has been deposited by a river which has lost its energy at the boundary between a steep slope and a gentle slope, for example when a mountain stream flows into a lake

**alluvium** the fine mud and silt deposited by rivers

**aquifer** a layer of rock which can hold water

**attrition** a form of erosion which is caused by particles and pebbles hitting each other

**basin lag** the time difference between the peak rainfall and the peak of a river's flow

**braiding** the division of a river into two or more streams after material has been deposited on the river bed

**corrasion** a form of erosion which is caused by particles scraping against the bed and sides of the river

**delta** the silt deposited at the mouth of a river; this happens after the river loses power and has to deposit its load

**deposition** the dropping of material which has been eroded and transported

**desertification** the process by which a desert gradually spreads into a neighbouring semi-desert area, for example the Sahara Desert in north Africa is spreading into the neighbouring Sahel region

**discharge** the quantity of water which flows past a fixed point in a river over a certain time

**drainage basin** the area drained by a river and its tributaries

**erosion** the wearing down and removal of rock

**evapotranspiration** the loss of moisture by evaporation from water surfaces and the soil and by transpiration from plants

**flood plain** the wide, flat plain formed when a river deposits its load in its lower course

**groundwater** all the water which seeps

underground into the spaces and pores in rocks; groundwater moves slowly to rivers and the oceans

**gully** a narrow, steep-sided channel on a hill or mountain slope which is formed by the action of water (usually because of rapid runoff after heavy rainfall)

**hydraulic action** a form of erosion which is caused by the force of moving water

**hydrograph** a graph which shows the record of channel flow in a river or stream

**hydrological cycle** the water cycle in which water evaporates from the sea, forms clouds, falls as precipitation and then drains back into the sea

**impermeable rock** a rock which water can't pass through

**infiltration** the process of water soaking into the ground; the infiltration rate is the speed at which water soaks into the ground

**interception** the ability of plants, and objects such as buildings and roads, to store precipitation water on their surfaces for a short time

**levees** natural embankments formed by the deposition of silt when a river floods

**meander** a sharp bend or curve in a river

**oxbow lake** a small crescent-shaped lake formed after a river wears away at the outside bend of a meander so that the neck of the meander becomes narrower and narrower until the river breaks through; the deposition of sediment results in the oxbow lake being cut off from the river

**permeable rock** a rock which water can pass through

**point bar** the deposition of sand and gravel on the inside of a meander loop

**porous rock** a rock with holes or pore spaces which enable it to hold water

**potholes** holes in river beds formed by the erosive power of pebbles being swirled around by the water (abrasion)

**regime** the pattern of seasonal variations in the volume of water in a river

**river cliff** a steep bank which is created when a river cuts into the side of its valley

**runoff** rainwater running off from where it falls

**Sahel** the southern edge of the Sahara Desert in north Africa which has experienced many famines because of long droughts

**soil erosion** the removal of soil from an area by means of water or wind; soil erosion is most likely to occur when the soil has been left bare, for example in the Sahel of north Africa

**terraces** the remains of former flood plains through which the river has cut

**throughflow** the flow of water down a slope through the soil; this usually happens when the amount of water falling on to the surface of the land is too great for it all to be absorbed quickly downwards

**transpiration** the process by which plants lose water

**transportation** the movement of material in rivers by suspension, bed load and solution

**watershed** the boundary of a drainage basin

**water table** the level at which all the pores in the rock are full of water; the height of the water table rises or falls according to recent rainfall

## TASKS

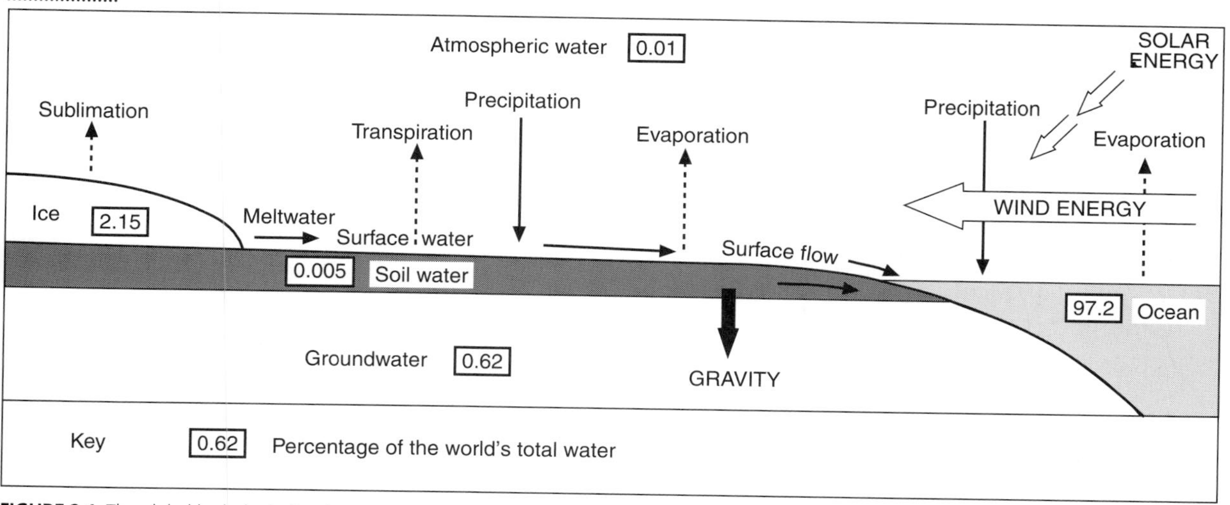


**FIGURE 2.1** The global hydrological cycle

**1** Study the diagram of the global hydrological cycle shown as Figure 2.1.

**(a)** Describe and explain the movement of water within the global hydrological cycle. (4)

**(b)** Explain how a balance is maintained within the hydrological cycle. (6)

**(10)**

**2** Figure 2.2A shows the rainfall variations in a Sahel region of north Africa. Figure 2.2B shows a newspaper article.

| Month | Year 1 | Year 2 | Year 3 | Year 4 | Year 5 |
|---|---|---|---|---|---|
| | Rainfall (mm) | | | | |
| January | — | — | — | — | — |
| February | — | — | — | — | — |
| March | — | 2 | — | — | 2 |
| April | 4 | — | 32 | — | — |
| May | 26 | — | 1 | 6 | — |
| June | 6 | 6 | 16 | 2 | 2 |
| July | 16 | 62 | 56 | 20 | 16 |
| August | 60 | 56 | 24 | 45 | 2 |
| September | 70 | 30 | 8 | — | — |
| October | 40 | — | — | — | — |
| November | 36 | — | — | — | — |
| December | — | — | — | — | — |
| Total | 258 | 156 | 137 | 73 | 22 |

**FIGURE 2.2A** Rainfall variations in a Sahel region

# Thousands Die in Sahel Drought

*Addis Ababa, Ethiopia*

Famine on a massive scale is causing thousands of deaths in the Sahel region of Ethiopia and Sudan. In some parts of the Sahel rain hasn't fallen for over a year and the land which was expected to offer crops of maize and millet has turned into a huge dustbowl. The scene today is a stark one with hundreds of thousands of people scratching the surface of the land looking for food .........

**FIGURE 2.2B** A newspaper article

**(a)** Describe the variations in rainfall shown by Figure 2.2A. (5)

**(b)** Outline the effects of rainfall variability on the people living in the Sahel. (5)

**(10)**

**3** Study Figure 2.3, which shows the basin hydrological cycle.

**(a)** Describe the inputs and outputs in the basin hydrological cycle. (3)

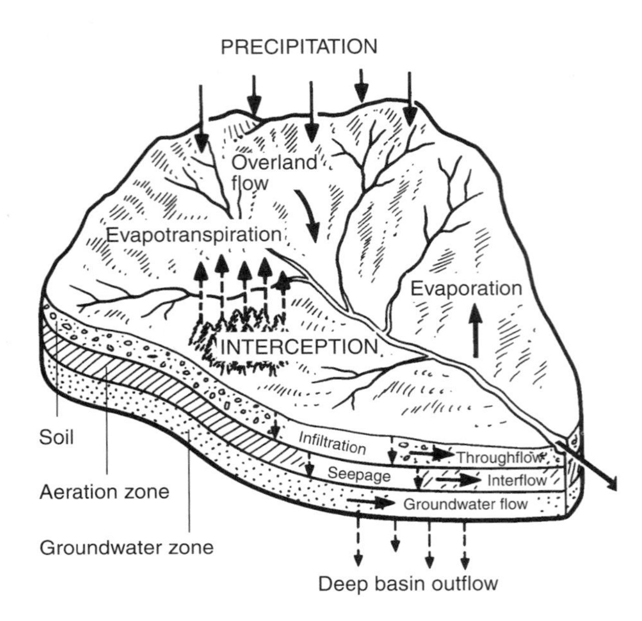


**FIGURE 2.3** The basin hydrological cycle

**(b)** State the meaning of the following terms:
i. interception;
ii. infiltration;
iii. throughflow. (3)

**(c)** Outline the ways people can affect the movement of water in the basin hydrological cycle. (4)

**(10)**

**4** Study Figure 2.4, which shows landscape features within a drainage basin.

**(a)** Name the features marked 1 to 8. (4)

**(b)** With the aid of diagrams, describe the processes which produced any two of the features marked 1 to 8. (6)

**(10)**

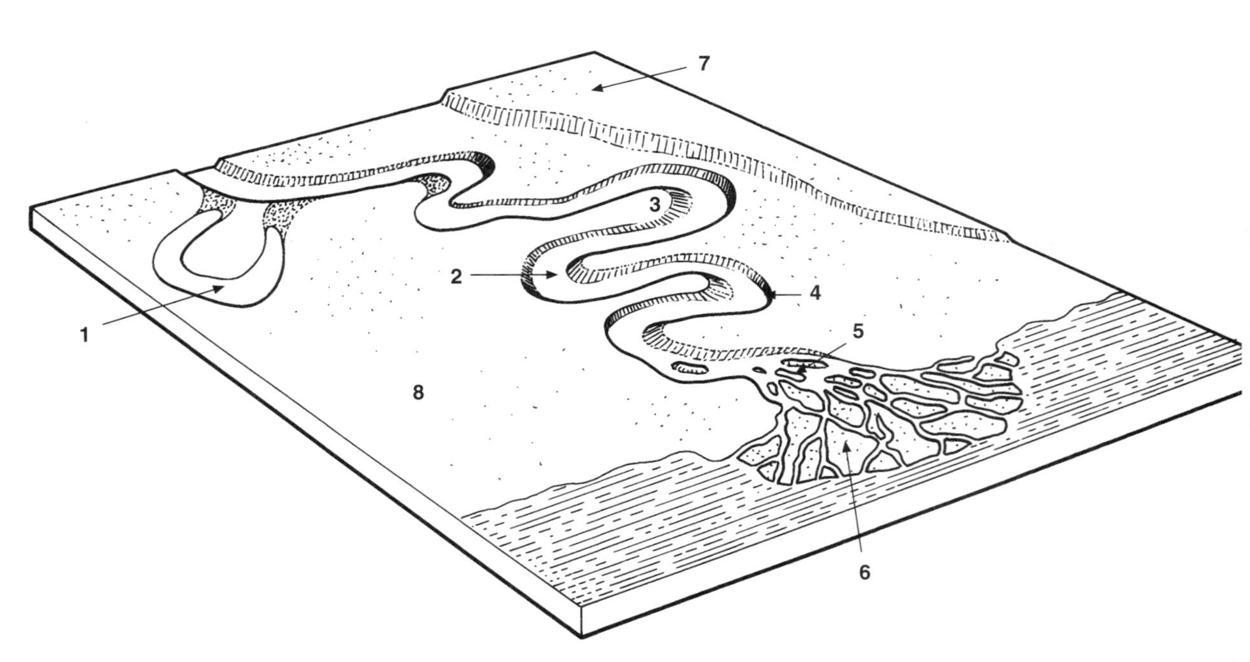


**FIGURE 2.4** Landscape features within a drainage basin

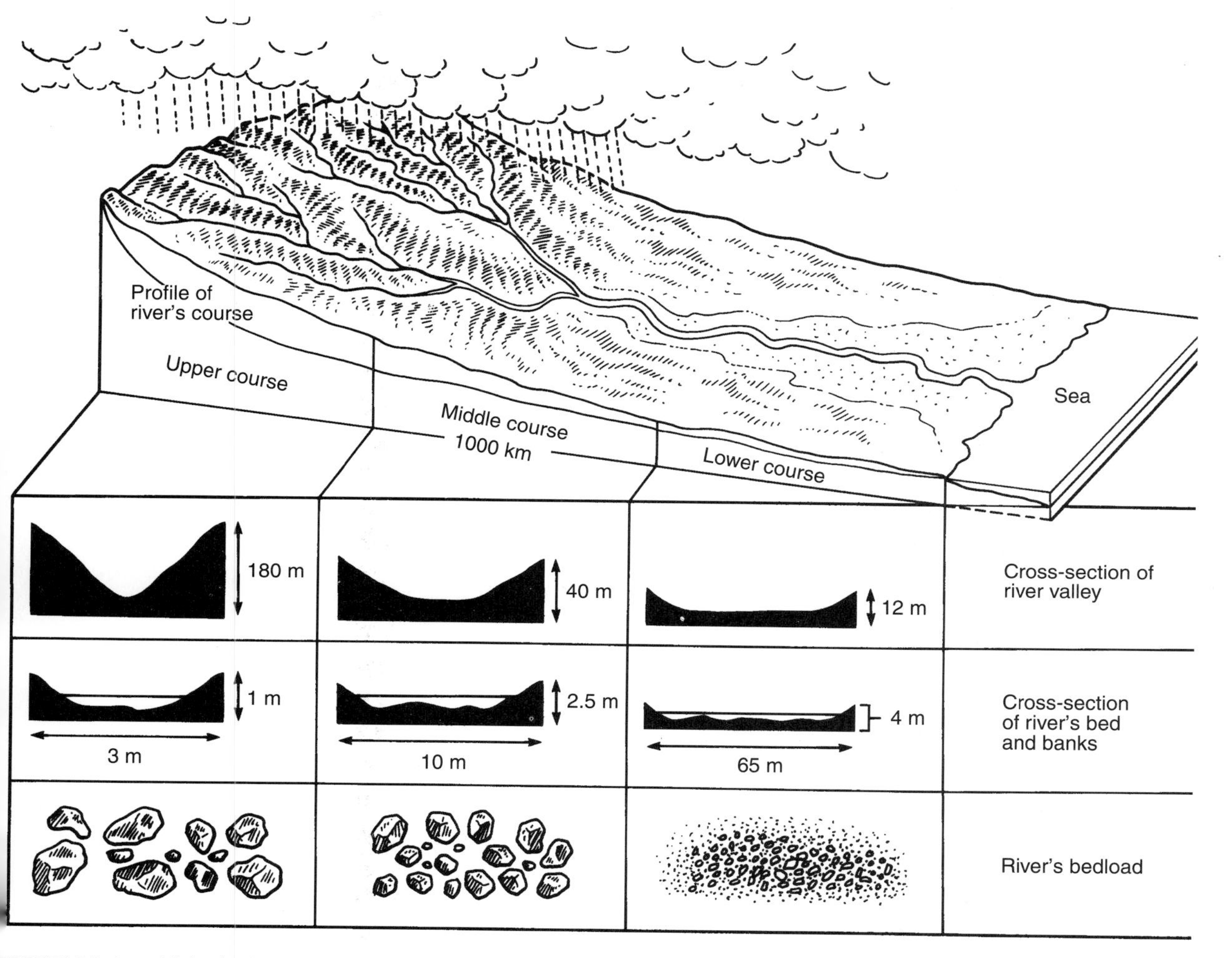


**FIGURE 2.5** A model river basin

**5** Figure 2.5 shows a model river basin. Describe and explain the different shapes of the river valley at its upper, middle and lower course. (10)

**(10)**

**6** Figure 2.6 shows the upper courses of rivers on a Scottish mountain.

**(a)** Describe the river courses shown in the photograph. (3)

**(b)** Outline the processes which shaped the river courses shown in the photograph. (7)

**(10)**

**FIGURE 2.6** The upper courses of rivers on a Scottish mountain

**7** Look at Figure 2.7A which shows rainfall and discharge figures for the section of the Endrick Water shown in Figure 2.7B.

**(a)** Outline the methods which would have been used to collect these figures. (3)

**(b)** Draw a hydrograph to show these figures. (3)

**(c)** Describe and explain the pattern shown by your completed hydrograph. (4)

**(10)**

| date | time | rainfall (mm) | discharge cubic metres/second |
|---|---|---|---|
| October 8 | 1600–1800 | 0 | 12 |
| | 1800–2000 | 6 | 10 |
| | 2000–2200 | 20 | 10 |
| | 2200–2400 | 45 | 16 |
| October 9 | 0000–0200 | 50 | 18 |
| | 0200–0400 | 50 | 22 |
| | 0400–0600 | 55 | 28 |
| | 0600–0800 | 50 | 35 |
| | 0800–1000 | 45 | 45 |
| | 1000–1200 | 20 | 60 |
| | 1200–1400 | 15 | 68 |
| | 1400–1600 | 5 | 60 |
| | 1600–1800 | 2 | 55 |
| | 1800–2000 | 1 | 50 |
| | 2000–2200 | 0 | 42 |
| | 2200–2400 | 0 | 30 |
| October 10 | 0000–0200 | 0 | 22 |
| | 0200–0400 | 0 | 12 |

**FIGURE 2.7A** Rainfall and discharge figures for a section of the Endrick Water

**FIGURE 2.7B** A section of the Endrick Water

**8** Examine the OS map on the inside back cover of this book.

**(a)** Describe the physical features of the section of the River Leven, and its tributaries, shown on the map. (5

**(b)** With the help of a diagram, account for the formation of one of the features you mentioned in (a) above. (5

**(10**

TOTAL NUMBER OF MARKS 8

# 3

# Lithosphere

## *Key topics*

- the processes of erosion, transportation and deposition;
- the influence of structure and rock type;
- glaciated upland landscapes – their formation and characteristic features;
- scarp and vale scenery – their formation and characteristic features;
- upland limestone landscapes – their formation and characteristic features;
- the identification of landscape features on OS maps, aerial photographs and sketches;
- the modification of relief forms by active geomorphological processes;
- the nature and effects of weathering;
- the effects of mass movements such as soil creep, slopewash, mudflows, scree slopes, landslips and landslides.

## *Key skills*

- constructing and interpreting cross-sections or transects;
- analysing statistical data on speed of movements, water content, stone characteristics and slope angle in mass movements;
- identifying landscape features on OS maps.

## *Key words*

**gents of erosion** ice, water, wind and the other means by which the land is eroded

**rête** a knife-edge ridge between two corries

**iotic weathering** the breakdown of rocks by the action of animals and plants

**arboniferous limestone scenery (or karst scenery)** scenery where the surface rocks have been weathered into clints and grykes, and where water sinks underground to form potholes and caverns; examples can be found in the Malham district of Yorkshire

**hemical weathering** the breakdown of rocks by chemical reactions such as those caused by rain containing carbon dioxide from the atmosphere

**lints** flat blocks of limestone which, along with grykes, form limestone pavements

**corrie** a large hollow which has been carved out of a mountain by ice

**escarpment (or cuesta)** a hill with a steep scarp slope and a more gentle dip slope; examples can be found in the North and South Downs in south east England

**exfoliation** a weathering process in which the surfaces of rocks flake off in concentric layers like the skin of an onion (it is also known as onion-skin weathering)

**freeze-thaw** the freezing and thawing of water in rocks which results in the rocks breaking up

**glacial abrasion** the scraping or wearing away of rocks by other rocks embedded in glaciers

**glacial erosion** the wearing down and removal of rocks by glaciers

**glacial plucking** a form of glacial erosion which involves glaciers freezing on to rocks and then wrenching them away

**glaciated uplands** upland areas with corries, arêtes, hanging valleys and U-shaped valleys which have been formed by glacial erosion; examples include the Lake District of north west England

**grykes** enlarged joints, which along with clints, make up limestone pavements

**hanging valley** a smaller valley, produced by a smaller glacier, hanging over a larger valley produced by a larger glacier

**landslide** the sudden movement of rock, soil and vegetation down a slope

**limestone pavement** an outcrop of carboniferous limestone in which horizontal rocks have been weathered into clints and grykes

**limestone pillars** columns of calcite formed when stalactites merge into stalacmites in limestone caverns

**lithosphere** the solid layer of rock which makes up the surface of the earth

**mass movement** the movement of rock and soil down a slope

**mechanical weathering** the disintegration of rocks mainly because of repeated heating and cooling; the heat of the sun causes the rocks to expand and then at night when the rocks cool down they contract; the repeated expansion and contraction causes rocks to crack

**moraine** the rocks, pebbles and clays deposited by glaciers; there are four types of moraine: lateral moraine along the sides of valleys, medial moraine at the junction of two valleys, terminal moraine at the end of the glacier and ground moraine left beneath the ice

**mudflow** the movement of soil which has been liquefied by rain or melting snow

**pyramidal peak** a horn peak which is formed after several corries have been carved out of the top of a mountain

**scree** loose rock formed by frost shattering on the hillside

**soil creep** the gradual movement of soil down a slope

**solution** a form of chemical weathering in which chemicals in rainwater dissolve elements in rocks such as limestone

**stalactite** deposits of calcite hanging from the roof of a limestone cave

**stalagmite** deposits of calcite building up from the floor of a limestone cave

**U-shaped valley** a flat-bottomed, steep-sided valley formed by glacial erosion

**weathering** the gradual breakdown of rocks caused by mechanical, chemical and biotic processes; no movement of the weathered material is involved (otherwise the process would be classified as erosion)

## TASKS

**1** Look at Figure 3.1, which shows a glaciated upland landscape.

**(a)** Name the landscape features marked 1 to 6. (3)

**(b)** Draw a labelled sketch to help you explain how any one of the features marked 1 to 6 was formed. (4)

**(c)** Describe ways in which the landscape shown as Figure 3.1 is being modified by post-glacial weathering and erosion. (3)

**(10)**

**2** Figure 3.2 shows a transect of a scarp and vale landscape in south east England.

**(a)** Describe in detail the main features of this type of landscape. (5)

**(b)** Outline the influence of rock type in the development of this type of landscape. (5)

**(10)**

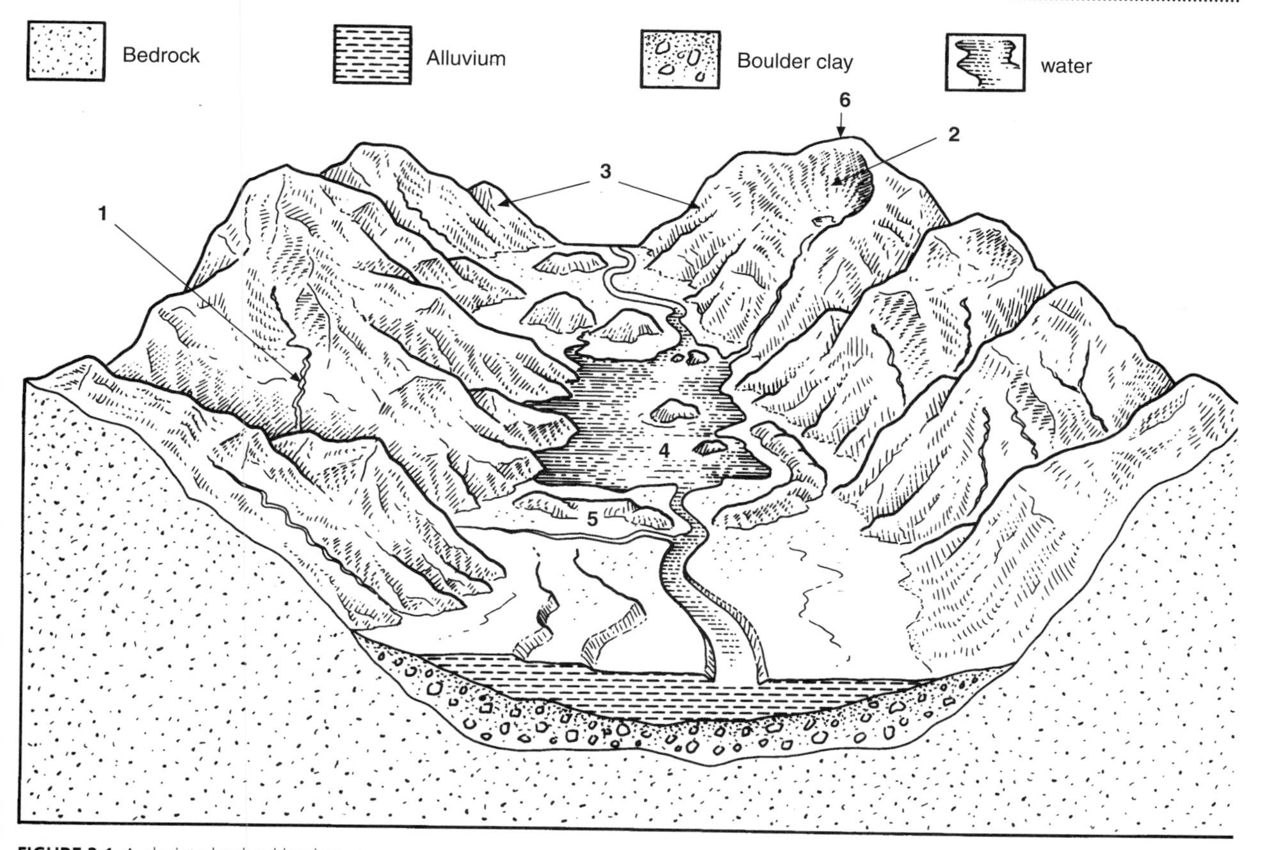


**FIGURE 3.1** A glaciated upland landscape

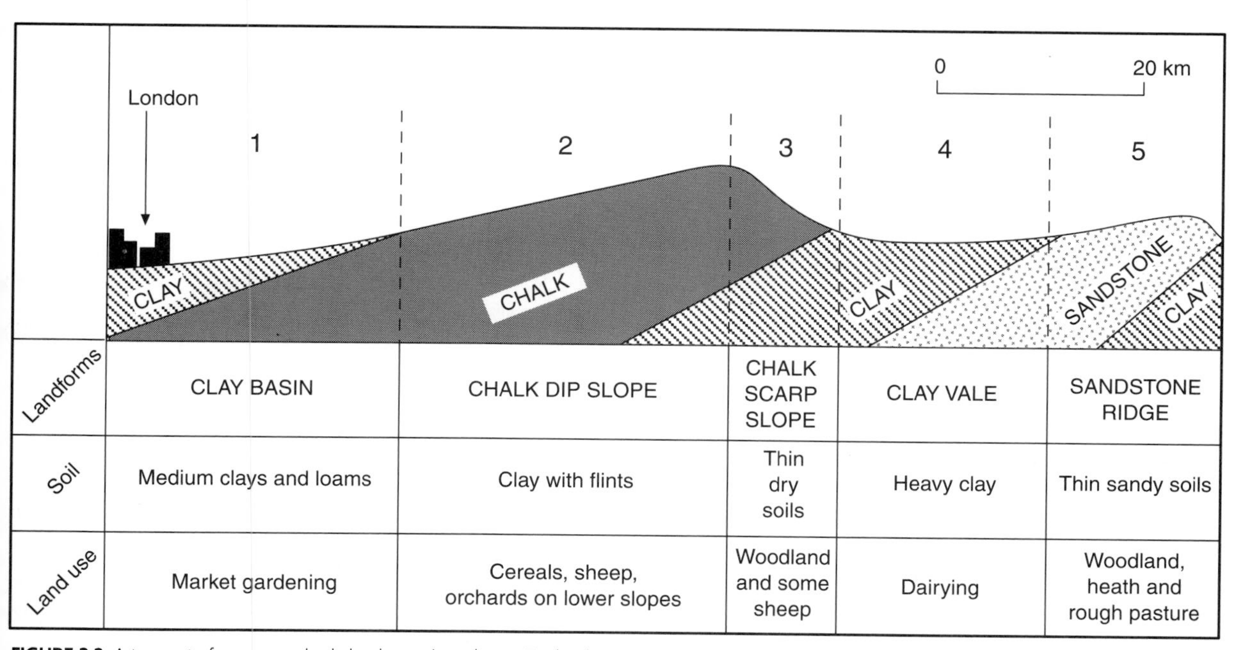


**FIGURE 3.2** A transect of a scarp and vale landscape in soth east England

**3** Look at Figure 3.3, which shows a carboniferous limestone area.

**(a)** Draw a labelled sketch to help you describe the main surface and underground features associated with this type of landscape. (5)

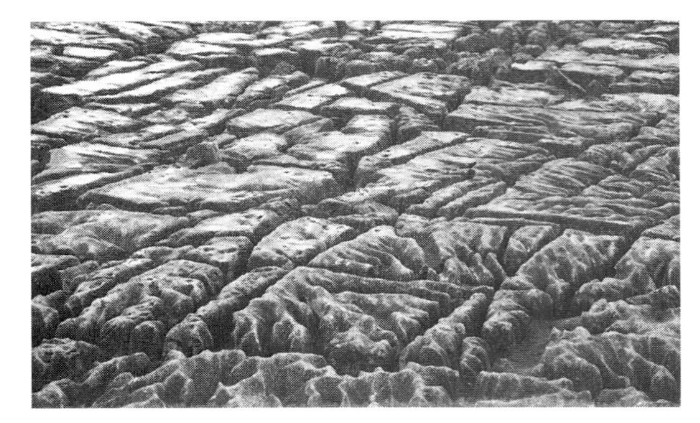

**FIGURE 3.3** An upland limestone area

**(b)** For any two of the features marked on your sketch, outline the processes which were involved in their formation. (5)

**(10)**

**4 (a)** Identify the landscape feature shown in Figure 3.4. Outline how this feature was formed. (4)

**(b)** With the help of examples, describe the difference between weathering and erosion. (3)

**(c)** State the meaning of the following terms:
i. mechanical weathering;
ii. biotic weathering;
iii. chemical weathering. (3)

**(10)**

**FIGURE 3.4**

**5** Figure 3.5A shows a classification of mass movements while Figure 3.5B is a newspaper article about a landslide in northern Japan.

**(a)** Describe the nature and effects of each of the mass movements shown in Figure 3.5A. (6)

**(b)** Outline the causes of landslides. (4)

**(10)**

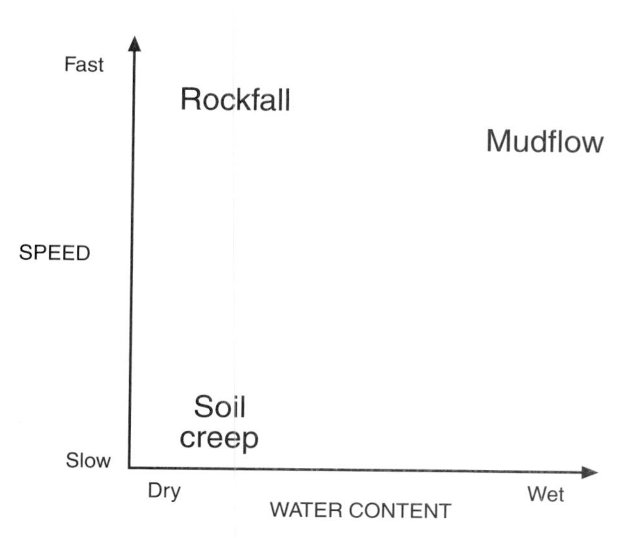


**FIGURE 3.5A** A classification of mass movements

**GIANT BOULDER BLOCKS TUNNEL**

*Furubira, Japan*

Rescue workers are working non-stop to remove a huge boulder which is blocking a road tunnel in Furubira, north Japan. Twenty people were left trapped in a coach and a car after the boulder crashed through the roof of the tunnel. The boulder, which is the size of 20-storey building, broke off the side of a mountain in a part of Japan which is regularly shaken by earth-quakes and .........

**FIGURE 3.5B** A newspaper article

**6** Look at Figure 3.6A and Figure 3.6B, which show two different types of weathering. Identify and describe the processes which are at work in each photograph. (10)

**(10)**

**FIGURE 3.6A**

**FIGURE 3.6B**

**FIGURE 3.7** An aerial photo of part of the Cairngorm Mountains

**7** Examine Figure 3.7, which shows an aerial photograph of part of the Cairngorm Mountains.

**(a)** Describe in detail the landform shown in this photograph. (5)

**(b)** Draw a labelled sketch to help you explain how the landform in the photograph was formed. (5)

**(10)**

**8** Look at the OS map of the Glencoe area on the inside back cover of this book.

**(a)** Draw a cross-section from the summit of Stob Coire nan Lochan (map reference 148548) to the summit of Sron Gharb (177585). (4)

**(b)** Write a detailed description of the landscape shown in your cross-section. (6)

**(10)**

TOTAL NUMBER OF MARKS **80**

4

# Biosphere

## Key topics

- the areal distribution of podzol, brown earth and gley soils;
- the properties of soils – their horizons, colour, texture and drainage;
- the influence of soil-forming processes on profiles – podzolisation, gleying, organic and nutrient movements;
- the evolution of coastal dune belts;
- the evolution of vegetation on derelict land;
- alterations to the vegetation of temperate deciduous and coniferous forests because of human interference.

## Key skills

- analysing soil profiles and data from soil surveys;
- interpreting and explaining data from vegetation surveys and distributions.

## Key words

**azonal soils** young soils which haven't changed much from their parent material

**biomass** the total weight of living organisms in a given area or ecosystem

**biome** a community of plants and animals which extends over a large area, for example the taiga biome or the hot desert biome

**biosphere** the part of the earth which can support life

**boreal forest** the coniferous forest also known as the taiga which lies to the immediate south of the tundra

**brown earth soils** the forest soils found in mid-latitudes

**chernozems** dark, humus-rich soils which are associated with mid-latitude grassland areas

**climax vegetation** the end point of a plant succession

**coniferous forests** forests of cone-bearing, usually evergreen trees such as pines, spruces and firs

**deciduous forests** forests of trees, such as oak and ash, which lose their leaves during the winter months

**dominant species** plants which, because of their height, control the environmental conditions of other plant species in the same community

**ecosystem** a community of living plants and animals and the environment in which they live

**flora** the plant species of an area

**fauna** the animal species of an area

**gleys** soils with too much water in them causing the oxides to become grey coloured

**horizons** the different bands of soil which make up a soil profile; the A-horizon is the topsoil, the B-horizon is the subsoil and the C-horizon is the bedrock

**humus** the dark mass of rotting organic material in soil

**intrazonal soils** soils which differ from zonal soils because of special conditions such as too much water

**iron pan** a layer of hard material found just below the surface of the ground; rainwater leaches chemicals including iron oxide, which sometimes solidifies below the surface of the ground
**leaching** the removal of chemicals and nutrients from the soil by, for example, rainwater; leached soils are usually coarse and infertile
**nutrients** the food substances which are required by plants and animals
**parent material** the bedrock from which a soil has been formed
**pedology** the study of soils
**permafrost** soil that is permanently frozen
**pioneer vegetation** the first plants to colonise an area
**plant community** a group of plants which share the same environment
**podzol soils** acidic soils which often occur in areas of coniferous forest
**sand dunes** ridges of sand
**soil profile** a vertical section of the soil which shows the different horizons or layers
**succession** the steady change of plants in an ecosystem, for example in temperate forest regions the succession begins with a simple pioneer community of grasses – this is followed by small plants which add organic matter to the top soil – this enables shrubs, and then pine trees, to grow – this enables hardwood trees such as oak and ash to flourish
**taiga** the coniferous forest which stretches over large parts of North America and Asia
**topsoil** the top layer or A-horizon of the soil profile
**zonal soils** soils which are largely determined by climate and vegetation

## TASKS

1 Study Figure 4.1, which shows a slope with three different soils.

(a) Describe the type of soils most likely to found at sites X, Y and Z. (6)

(b) For any one of these soils, outline the factors which influenced its development. (4)

**(10)**

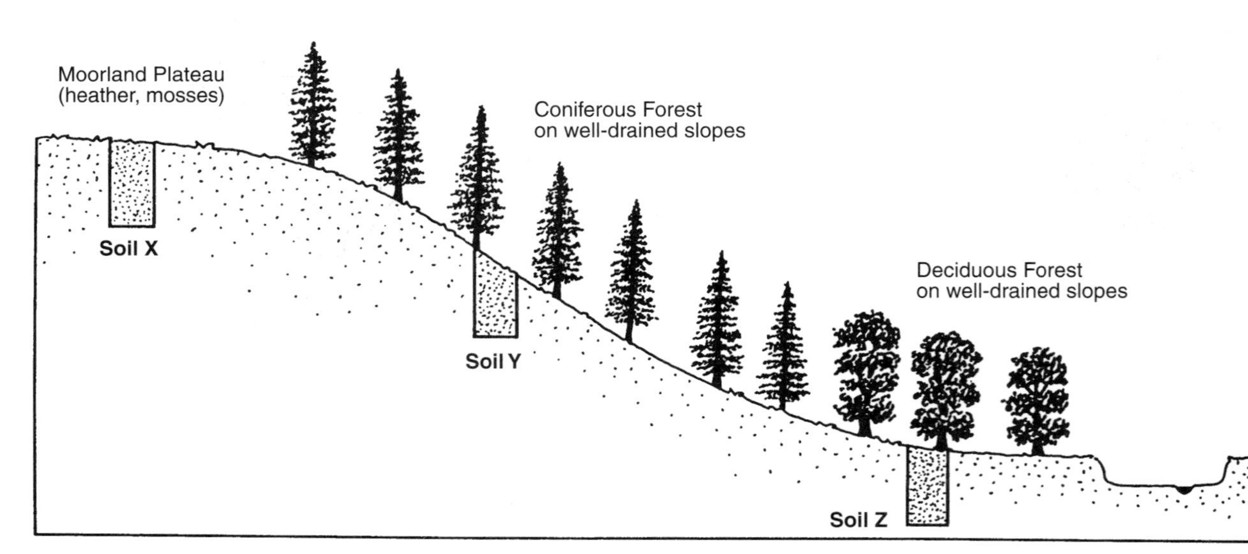


**FIGURE 4.1** A slope with three different soils

**2** Look at Figure 4.2, which shows two soil profiles. Identify and describe both soils, highlighting the features which helped you identify them. (10)

**(10)**

**3** Study the map of the distribution of podzols and brown forest soils shown as Figure 4.3.

**(a)** Describe the distribution of podzols and brown forest soils. (4)

**(b)** Outline the role of climate in the development of each of these soils. (6)

**(10)**

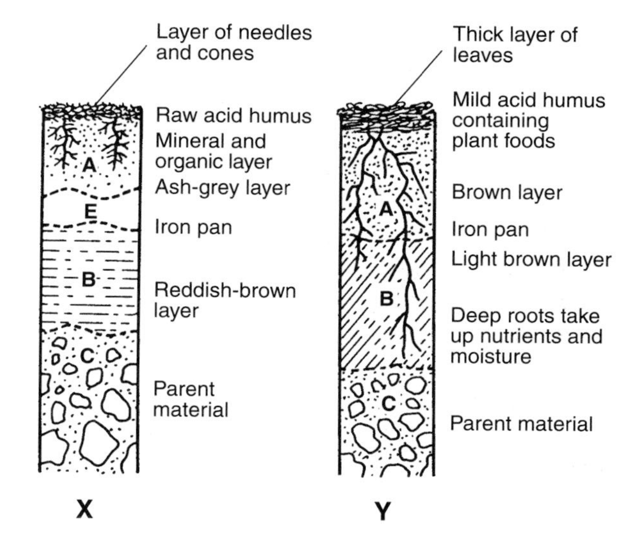


**FIGURE 4.2** Two soil profiles

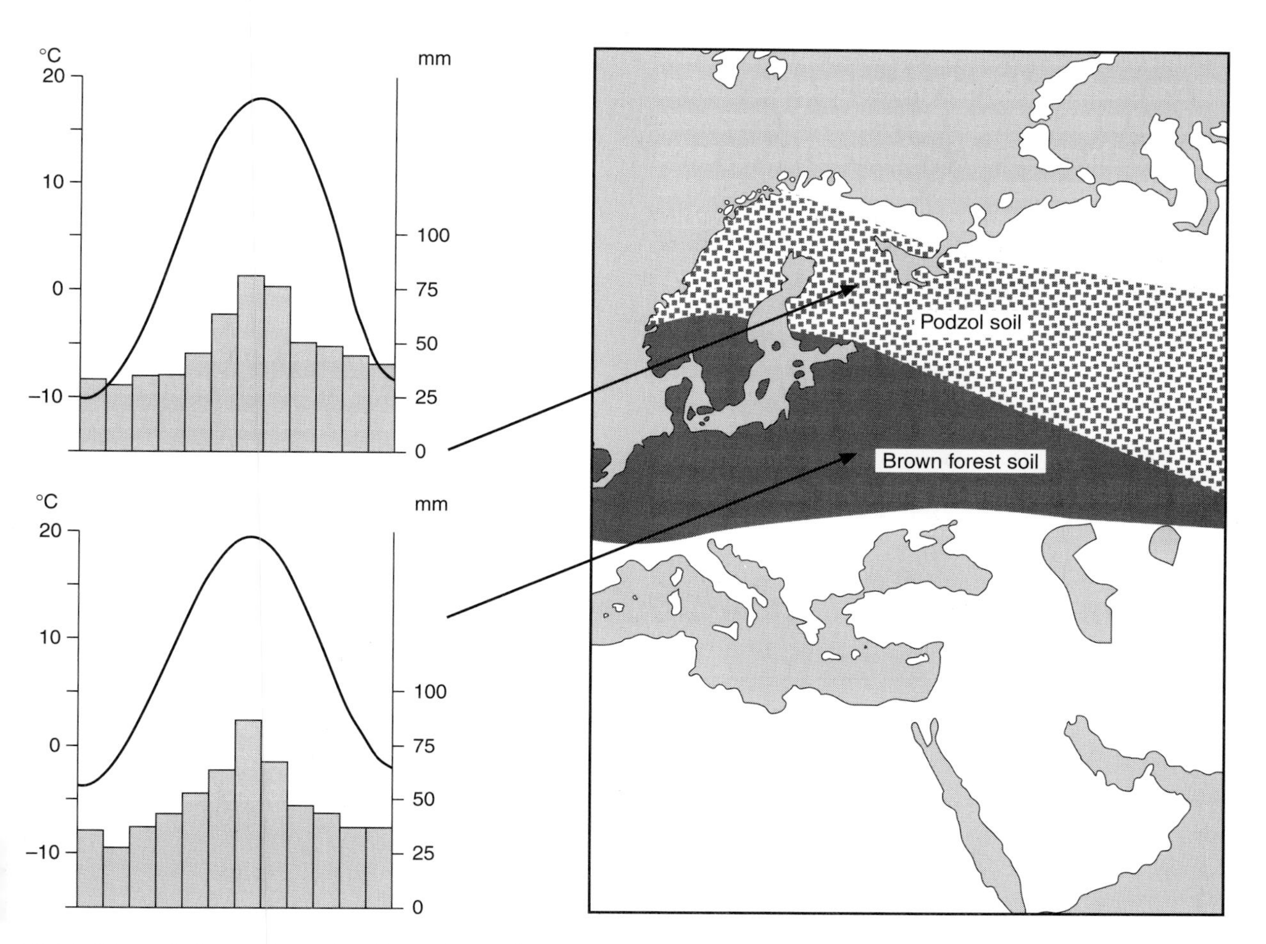


**FIGURE 4.3** The distribution of podzols and brown forest soils

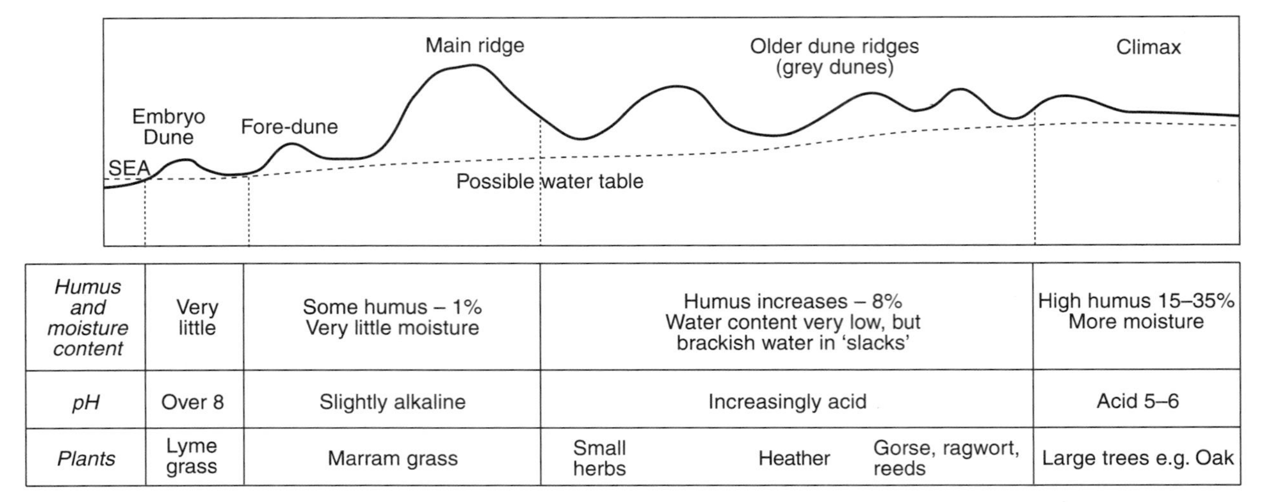


| Humus and moisture content | Very little | Some humus – 1% Very little moisture | Humus increases – 8% Water content very low, but brackish water in 'slacks' | | | High humus 15–35% More moisture |
|---|---|---|---|---|---|---|
| pH | Over 8 | Slightly alkaline | Increasingly acid | | | Acid 5–6 |
| Plants | Lyme grass | Marram grass | Small herbs | Heather | Gorse, ragwort, reeds | Large trees e.g. Oak |

**FIGURE 4.4** A transect across sand dunes

**4** Figure 4.4 shows a transect across sand dunes. Describe and account for the changes in plant type along the transect. (10)

**(10)**

**5** Examine Figure 4.5, which shows a typical plant succession in the temperate forest zone of North America.

**(a)** Describe and explain the plant succession shown as Figure 4.5. (5)

**(b)** State the meaning of the following terms:

i. plant community;
ii. pioneer vegetation;
iii. plant succession;
iv. climax vegetation;
v. dominant species. (5)

**(10)**

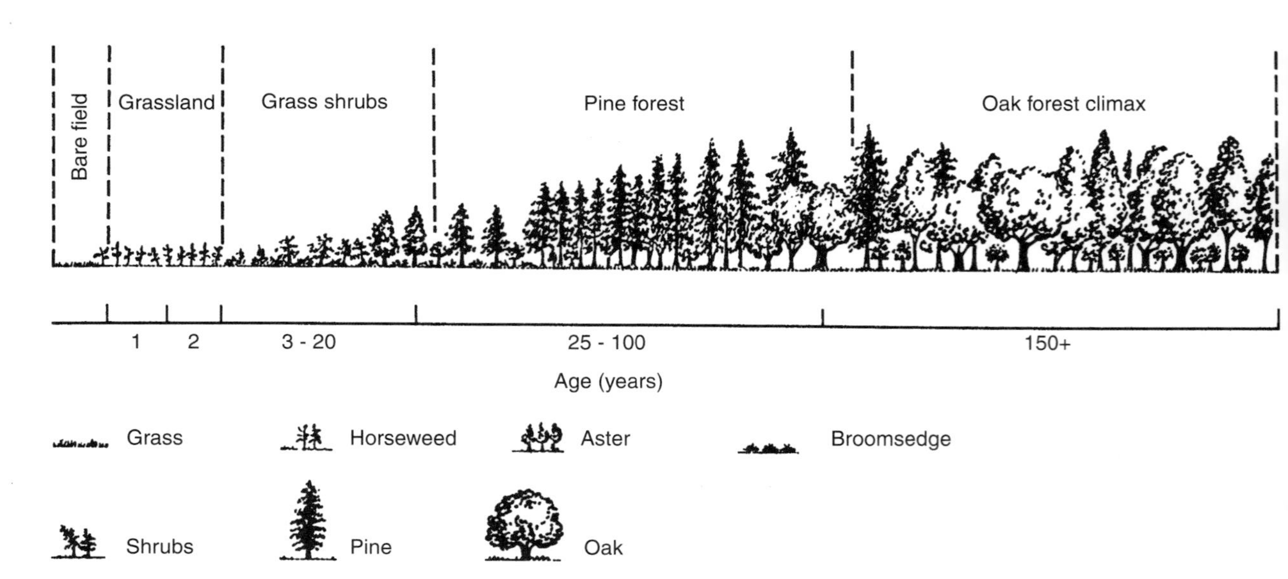


**FIGURE 4.5** A typical plant succession in the temperature forest zone of North America

**6** Look at Figure 4.6, which shows plants growing in an area of derelict land.

**(a)** Describe the types of plants most likely to be found at this site. (6)

**(b)** Outline the special conditions which affect the types of plants growing in derelict areas. (4)

**(10)**

**7** Study Figure 4.7, which shows the world distribution of coniferous forests and temperate deciduous forests.

**(a)** Describe and give reasons for the different types of trees which are associated with each of these forests. (5)

**(b)** Explain why temperate deciduous forests have been cleared at a much greater rate than coniferous forests. (5)

**(10)**

**FIGURE 4.6** Plants in an area of derelict land

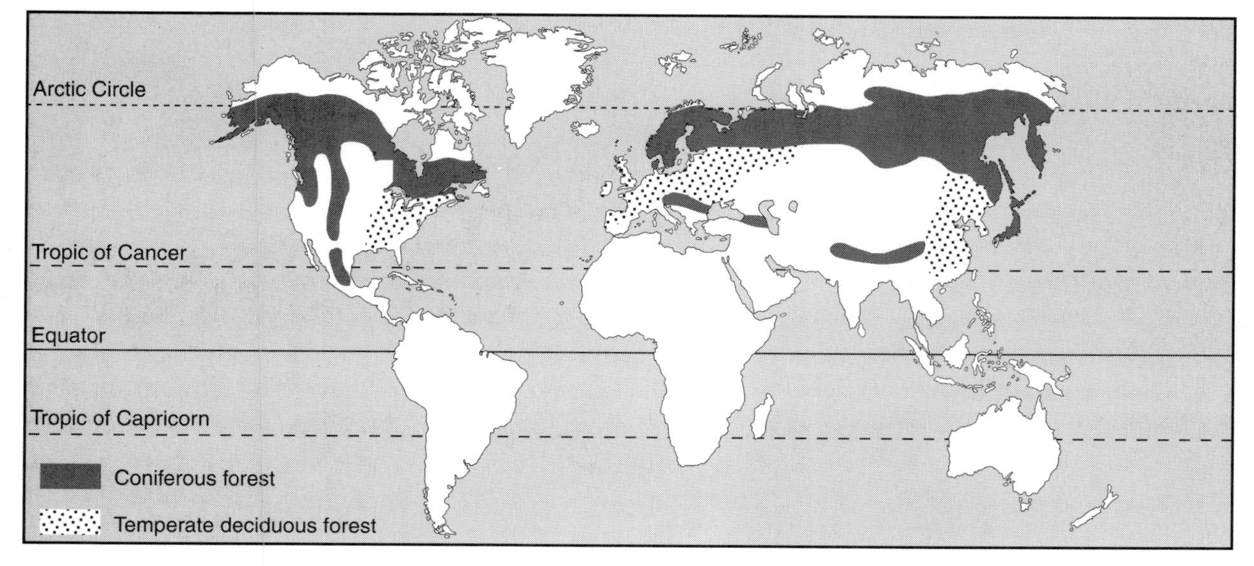


**FIGURE 4.7** The world distribution of coniferous forests and temperate deciduous forests

8 Figure 4.8A shows the results of a vegetation survey in a lowland area of northern Scotland. Figure 4.8B shows damage to vegetation in this area.

**(a)** Comment on the results of the vegetation survey shown as Figure 4.8A. (3)

**(b)** What are the consequences of the removal of trees for the other plants in an area? (3)

**(c)** Outline the ways in which human interference can damage an area's vegetation cover. (4)

**(10)**

| year | number of trees (over 3 metres high) |
|---|---|
| 1900 | 2500 (estimated) |
| 1910 | 2600 (estimated) |
| 1920 | 850 |
| 1930 | 1500 |
| 1940 | 1200 |
| 1950 | 850 |
| 1960 | 220 |
| 1970 | 110 |
| 1980 | 80 |
| 1990 | 25 |

**FIGURE 4.8A** The results of a vegetation survey in a lowland area of northern

**FIGURE 4.8B** Damage to vegetation

TOTAL NUMBER OF MARKS **80**

# 5 Population Geography

## Key topics

- population structures for developed countries and developing countries;
- censuses;
- birth rates, death rates, natural increase and the demographic transition model;
- how fertility, mortality and migration affect population change;
- the reasons for population change, including the factors which affect birth rates and death rates;
- the implications of population change;
- problems caused by high birth rates in developing countries;
- problems caused by ageing populations in developed countries;
- different types of migration;
- reasons for migration;
- a case-study of international migration (for example: between EU countries);
- a case-study of rural–urban and urban–rural migration (for example: within the UK);
- a case-study of forced migration (for example: refugee movements in the Sahel).

## Key skills

- interpreting population information on maps, diagrams or tables of data;
- interpreting flow diagrams.

## Key words

**active population** the number of people of working age

**ageing population** a population with a low birth rate and a low death rate so that the average age of the population is increasing

**census** a count of the number of people in a country

**counterurbanisation** the movement of people away from towns and cities

**crude birth rate** the number of live births per 1000 people in a year

**crude death rate** the number of deaths per 1000 people in a year

**demographic transition model** a model which shows how changes in birth rates and death rates are related to population growth over a long period of time

**dependent population** the number of people of non-working age

**depopulation** the movement of people away from an area

**developed countries** rich countries, such as the USA and Japan, with developed economies

**developing countries** poor countries, such as Bangladesh and Ethiopia, with economies which are still developing

**emigration** the movement of people out of a country

**fertility ratio** the ratio of children to women

**forced migration** the compulsory movement of people away from an area

**immigration** the movement of people into a country

**infant mortality rate** the number of deaths of infants under the age of one year (per 1000 live births)

**inter-continental migration** the movement of people from one continent to another, for example the movement of the 'boat people' from Vietnam to Australasia, Europe and North America

**inter-regional migration** the movement of people from one region of a country to another region in the same country, for example from rural areas in the Highlands of Scotland to urban areas in the Lowlands of Scotland

**international migration** the movement of people from one country to another country, for example from Scotland to Canada

**life expectancy** the average number of years a new-born baby can expect to live

**migration** the movement of people away from their homes

**mortality rate** the number of deaths per 1000 people

**natural increase** the birth rate minus the death rate

**overpopulation** a population which is bigger than that which the available resources can support

**population density** the number of people relative to the space or area in which they live (the number of people per square kilometre of land for example)

**population distribution** the location or spread of population

**population pyramid** a graph which shows the age and sex structure of a population

**population structure** the number of males and females in each age group

**pull factors** factors, such as jobs, better housing and good social facilities, which attract people to an area

**push factors** factors, such as natural disasters, unemployment and housing shortages, which force people to leave an area

**refugees** people who are forced to leave their home areas because of, for example, a war or famine

**rural-urban migration** the movement of people from the countryside to towns and cities

**urban-rural migration** the movement of people from towns and cities to the countryside

**seasonal migration** a short term movement of people, for example the movement of farm labourers to farming areas during the busy harvest period

**transmigration** the relocation of a large number of people away from overcrowded core regions to the less crowded outer regions; in Indonesia the government has relocated people from the overcrowded island of Java to Sumatra and other less populated islands

# TASKS

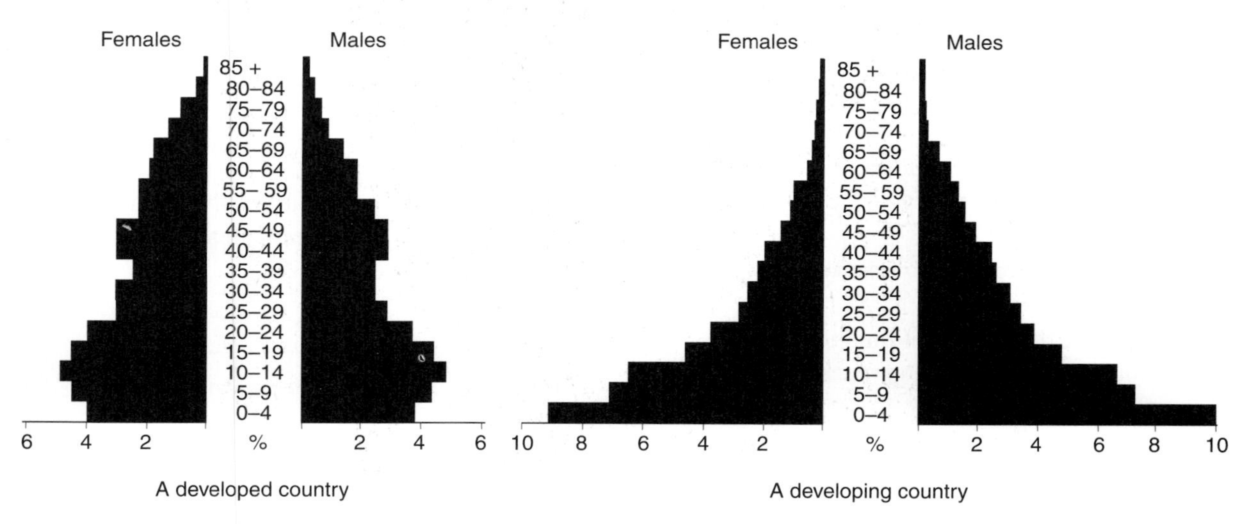


**FIGURE 5.1** A population pyramid for a Developed Country and a population pyramid for a Developing Country

**1** Figure 5.1 shows two population pyramids for a developed country and a developing country.

**(a)** State the meaning of the terms 'active population'and 'dependent population'. (4)

**(b)** Describe and account for the differences between the active and dependent populations of the two population pyramids shown as Figure 5.1. (6)

**(10)**

**2** Study Figure 5.2, which shows population change in a developed country and a developing country.

**(a)** For any named developed country or developing country, describe and explain the changes in the birth rate and death rate. (4)

**(b)** Discuss the implications of high birth rates in developing countries. (3)

**(c)** Discuss the implications of low birth rates in developed countries. (3)

**(10)**

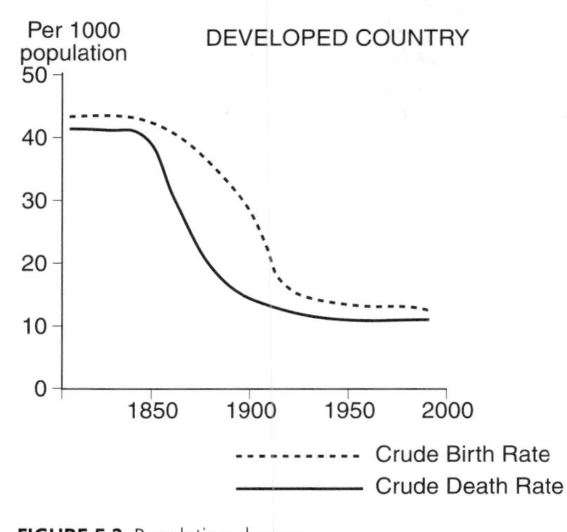


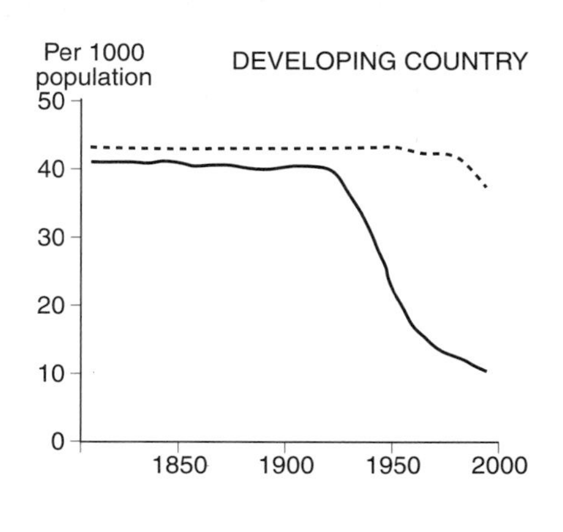


**FIGURE 5.2** Population change

**3** Figure 5.3 is a statement about census information.

> ***'The information collected from a census is crucial for effective planning, but not all countries are in a position to collect this information.'***

**FIGURE 5.3** A statement about census information

**(a)** Describe the type of information which would be gained from a properly-taken census. (4)

**(b)** Explain, with the help of examples, why some countries find it difficult to take a census. (4)

**(c)** Suggest ways in which the traditional way of life of the shifting cultivators can be protected. (2)

**(10)**

**4** Examine Figure 5.4, which shows a model of demographic transition. Describe and explain the trends shown in each of the model's four stages. (10)

**(10)**

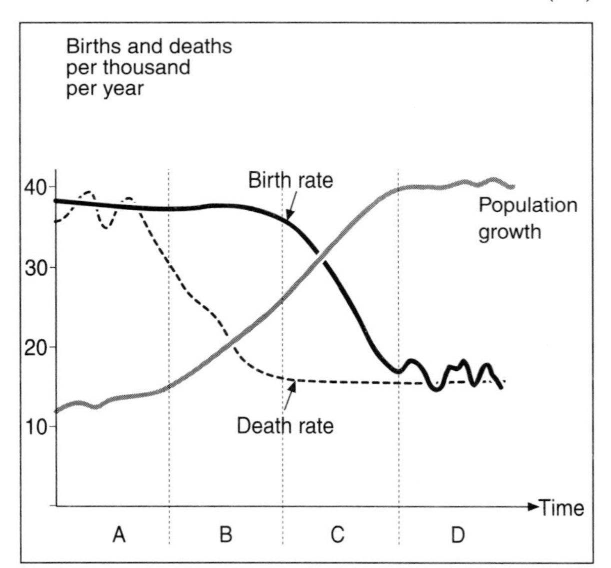


**FIGURE 5.4** A model of demographic transition

**5** Look at Figure 5.5, which shows a model of migration.

**(a)** Use the model to help you explain the migration of people between rural and urban areas in a developed country you have studied. (6)

**(b)** What problems may arise when large numbers of people leave an area? (4)

**(10)**

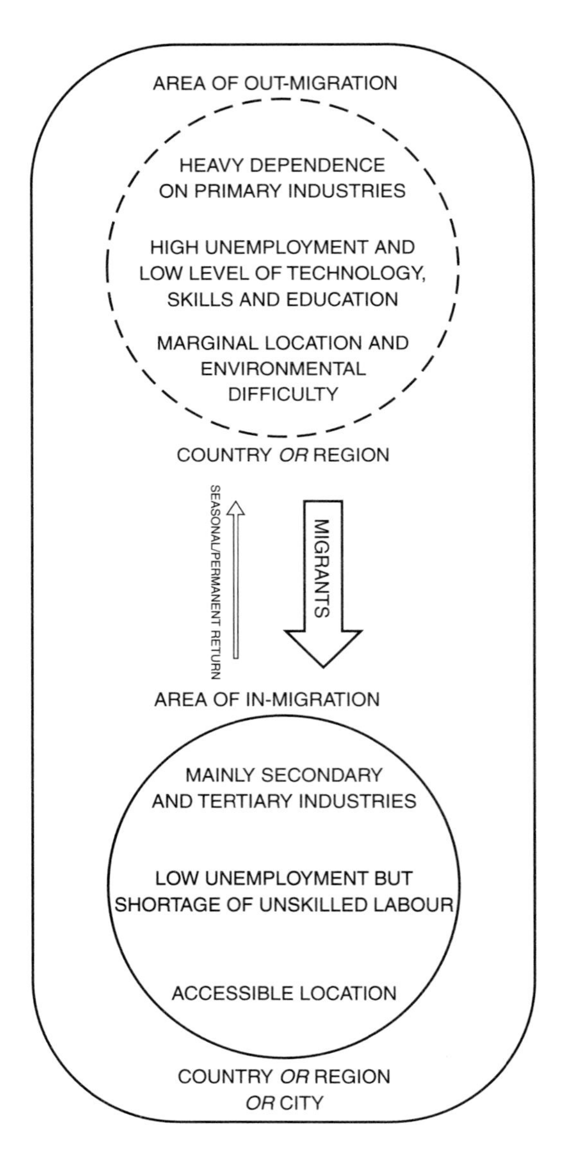


**FIGURE 5.5** A model of migration

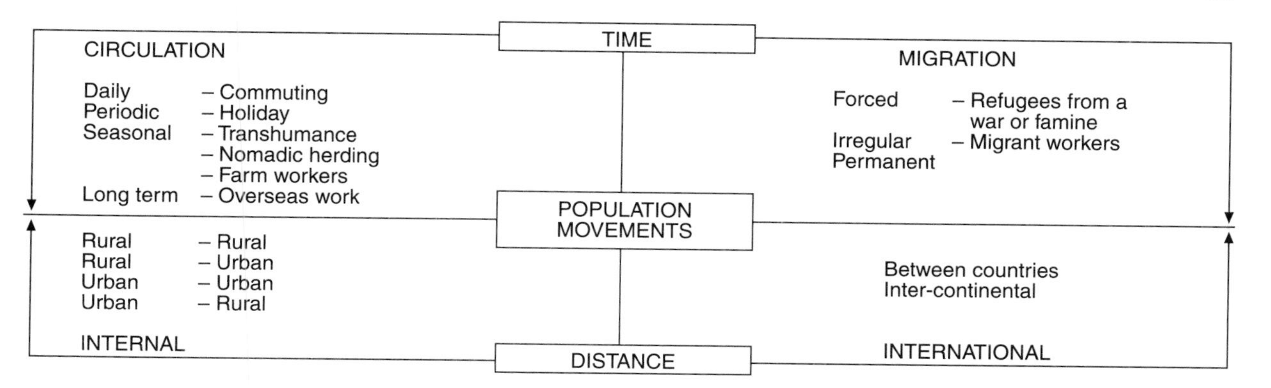


**FIGURE 5.6** A classification of population movement

**6** Figure 5.6 shows a classification of population movement.

**(a)** With reference to examples from both the Developed World and the Developing World, describe and explain refugee movements. (6)

**(b)** Describe how international agencies such as the United Nations can help to reduce the number of refugee movements. (4)

**(10)**

**7** Examine Figure 5.7, which shows the movement of migrant workers to West Germany during the 1970s.

**(a)** Outline the 'push' and 'pull' factors which encouraged the migration of workers to West Germany during the 1970s. (6)

**(b)** What were the benefits of this movement of migrant workers for West Germany? (4)

**(10)**

N
N = Netherlands
B = Belgium
S = Switzerland
A = Austria
Migrants × 1000
200
100
0
<1000
UK
N
B
France
S
A
Portugal
Spain
Italy
former Yugoslavia
Turkey
Morocco
Algeria
Tunisia
Greece

**FIGURE 5.7** The movement of migrant workers to West Germany during the 1970s

**8** Figure 5.8 shows population change in eight countries.

**(a)** State the meaning of the following terms:
   i. crude birth rate;
   ii. crude death rate;
   iii. natural increase;
   iv. overpopulation. (4)

**(b)** Suggest reasons for the differences between the natural population increases of different countries. (6)

**(10)**

| country | natural population increase (per 1000) |
|---|---|
| Tanzania | 33 |
| Bangladesh | 24 |
| India | 20 |
| Argentina | 13 |
| Canada | 7 |
| France | 3 |
| UK | 1 |
| Italy | 0 |

**FIGURE 5.8** Population change in eight countries

TOTAL NUMBER OF MARKS **80**

# 6 Rural Geography

## Key topics

- different types of agricultural systems;
- a case-study of shifting cultivation (for example: Amazonia);
- a case-study of intensive subsistence (peasant) farming (for example: the Ganges Valley);
- a case-study of extensive commercial farming (for example: East Anglia);
- different types of rural landscapes, including the field and infrastructure patterns associated with the case-studies listed above;
- rural change, including changes to the agricultural systems and rural landscapes listed above (for example: the impact of deforestation on shifting cultivation, the impact of the 'Green Revolution' on intensive peasant farming, and the impact of government and EU policy on extensive commercial farming;
- reasons for rural change;
- implications of rural change.

## Key skills

- analysing land use data and crop yields from maps, diagrams and tables of data;
- analysing farm survey results;
- annotating and analysing field sketches of rural landscapes.

## Key words

**agricultural system** the inputs, processes and outputs associated with farming
**cereal farming** the cultivation of cereals such as wheat and barley
**commercial farming** farming for money
**economic influences (on farming)** influences such as capital (money), transport, markets and the availability of government grants and subsidies
**extensive commercial farming** large-scale farming for money, for example the huge wheat farms on the prairies of Canada;
**Green Revolution** the farming improvements, including new seeds, fertilisers and equipment, which have enabled farmers in developing countries to produce more food
**human influences (on farming)** influences, including economic and technological influences, such as the availability of markets and the use of machines
**inputs** the land, seeds, fertilisers, machines, workers, etc. which are required for farming
**intensive subsistence (peasant) farming areas** the heavily-farmed areas of India, Bangladesh and other countries in south Asia
**outputs** the crops and other products which are produced by farms
**physical influences (on farming)** influences such as the shape and height of the land, the soil and the climate
**Prairies** the large plains of North America which are mainly used for growing cereals

**processes (in farming)** the sowing, fertilising, irrigating, harvesting, feeding and milking which is necessary to produce farm products

**relief** the shape and height of the land

**sedentary cultivation** cultivation which involves remaining in the same place

**set-aside land** land which has been taken out of farming in order to reduce the over-production of crops; farmers in the EU receive financial compensation for each hectare of land they take out of production and are allowed to use their set-aside land for non-farming purposes such as camp-sites, riding schools and nature reserves

**shifting cultivation** farming which involves moving from one area to another area, for example the shifting agriculture practised by the Indian tribes of the Amazon rainforest

**slash-burn farming** another name for shifting cultivation which involves cutting away the forest and burning the roots of the trees

**subsistence (peasant) farming** farming which usually only produces enough food for the farmer and his family to consume

**technological influences (on farming)** influences such as the development of new machines, fertilisers and seeds

## TASKS

**1** Study Figure 6.1, which shows the distribution of three agricultural systems.

**(a)** Describe the distribution of the three agricultural systems. (5)

**(b)** For each of the agricultural systems shown in Figure 6.1, describe the main features of the type of agriculture being practised. (5)

**(10)**

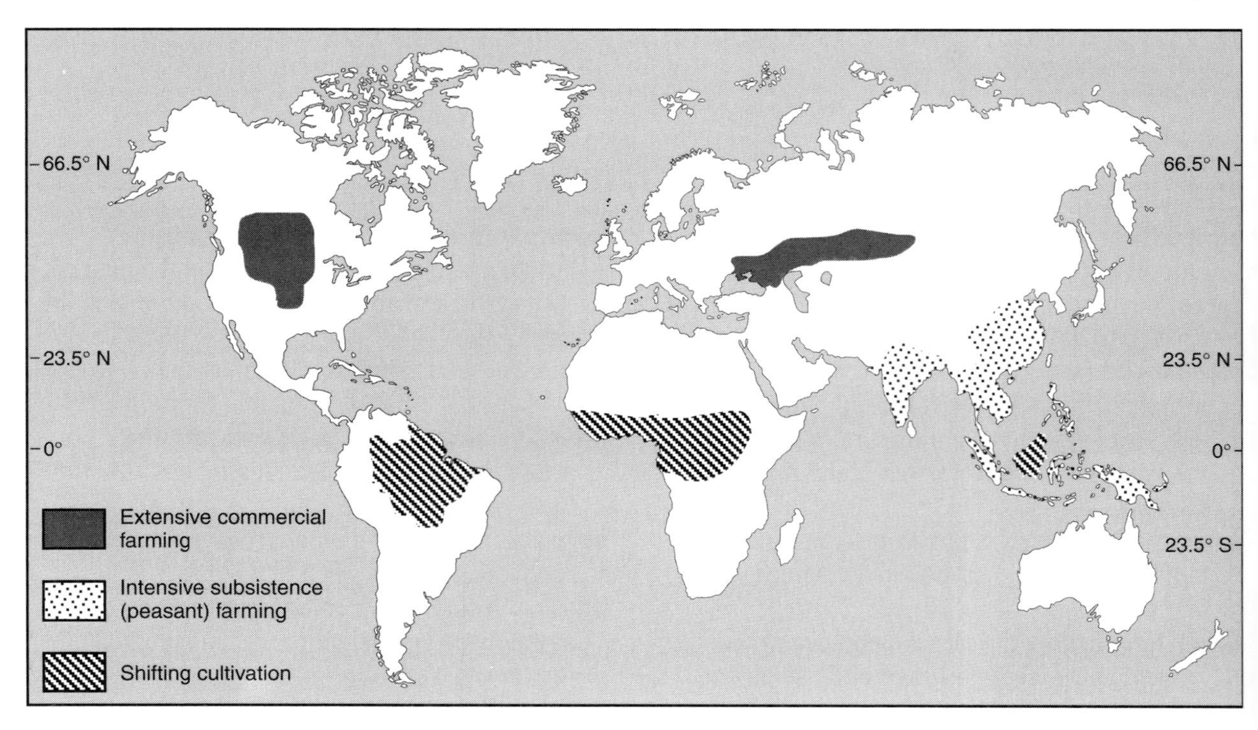


**FIGURE 6.1** The distribution of three agricultural systems

**2** Figure 6.2 is a statement about shifting cultivation.

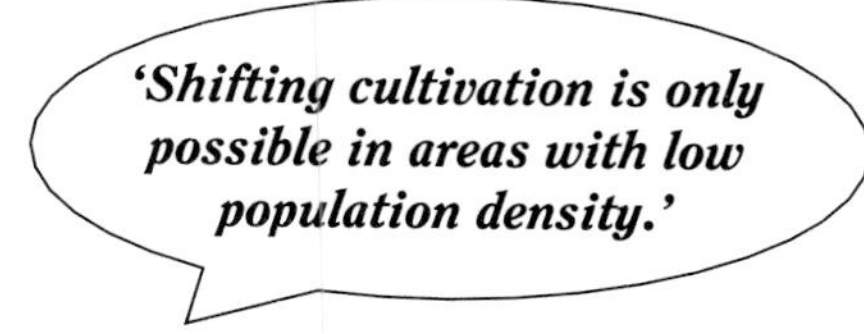


**FIGURE 6.2** A statement about shifting cultivation

**(a)** Explain why shifting cultivation is found in areas of low population density. (5)

**(b)** Areas of extensive commercial farming are also found in areas of low population density. Explain why. (5)

**(10)**

**3** Look at Figure 6.3, which shows changes in shifting cultivation in the Amazon rainforest.

**(a)** Describe and account for the changes shown by Figure 6.3. (4)

**(b)** Deforestation has had a major impact on shifting cultivation in the Amazon rainforest. Outline the ways in which deforestation has affected the shifting cultivators' way of life as well as the landscape in which they live. (4)

**(c)** Suggest ways in which the traditional way of life of the shifting cultivators can be protected. (2)

**(10)**

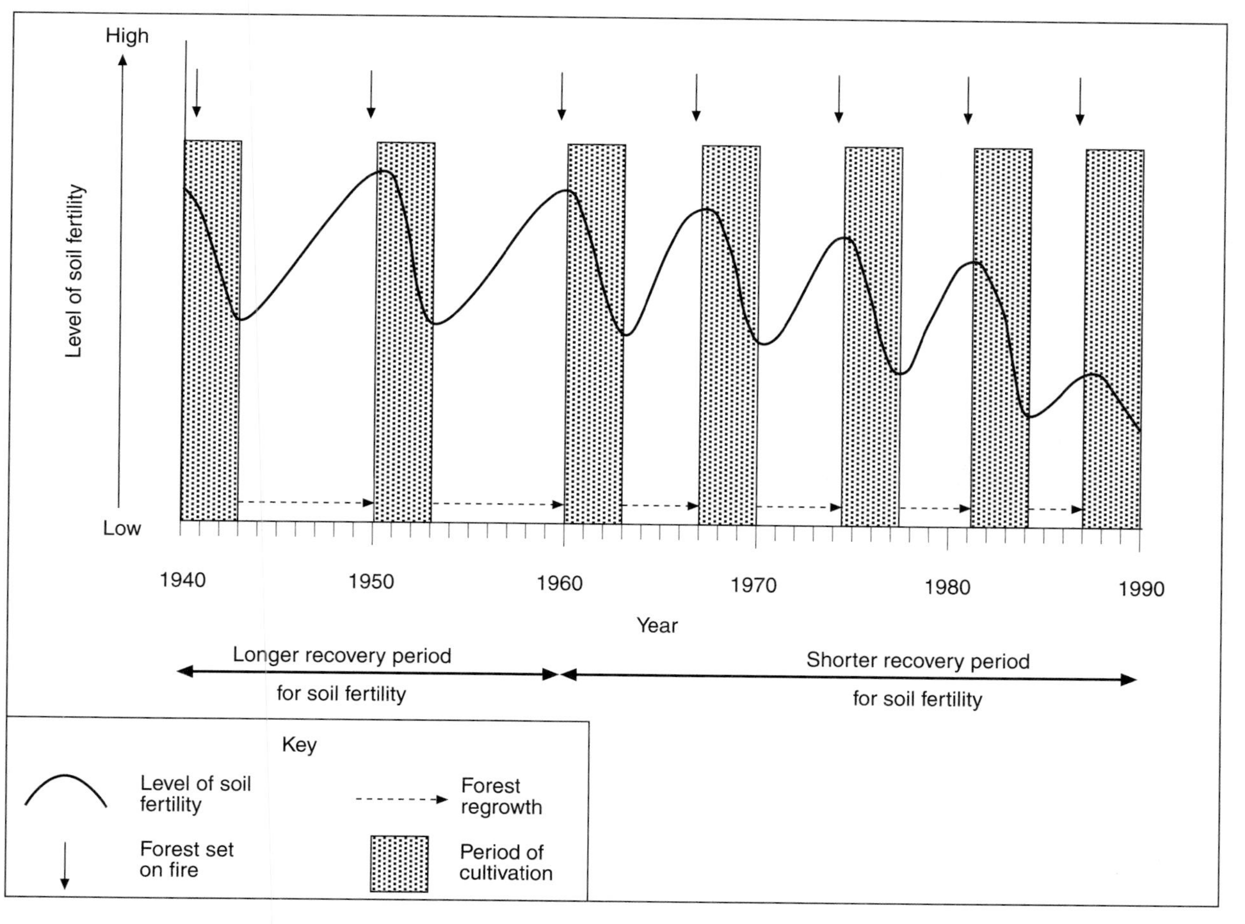


**FIGURE 6.3** Changes in shifting cultivation in the Amazon rain-forest

**4** Study Figure 6.4, which shows changes to a farm in East Anglia.

**(a)** Describe and explain the changes which have taken place on this farm. (6)

**(b)** State the meaning of the term 'set-aside land' and outline the advantages and disadvantages of set-aside land for the farmer concerned. (4)

**(10)**

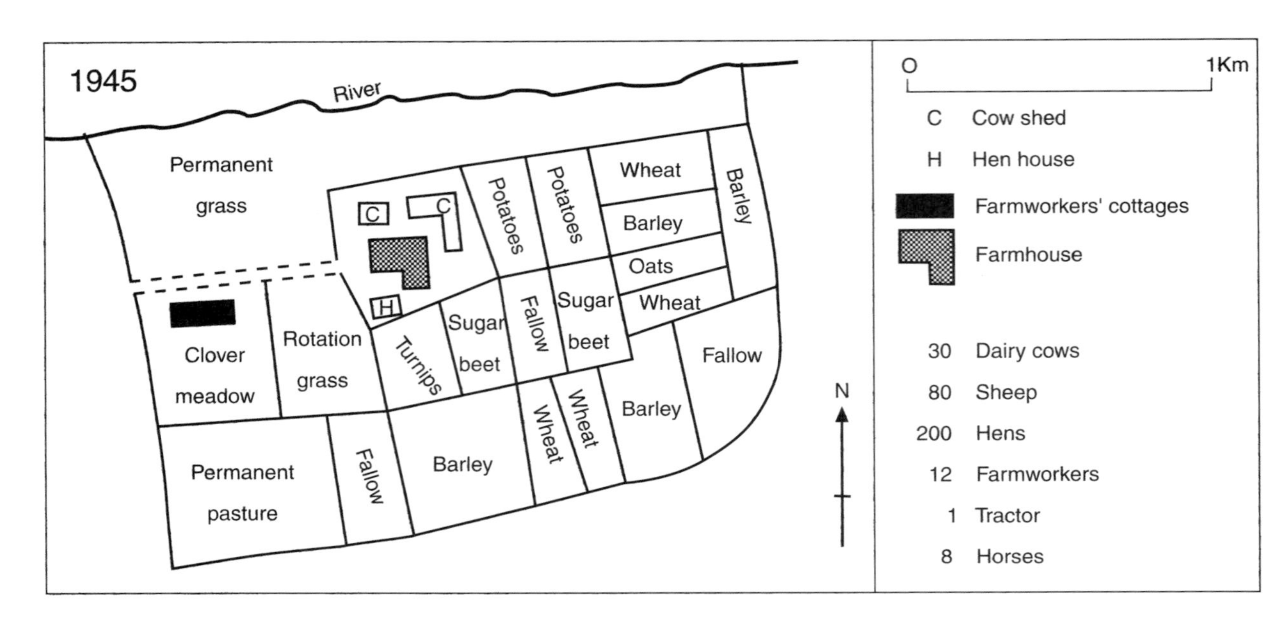


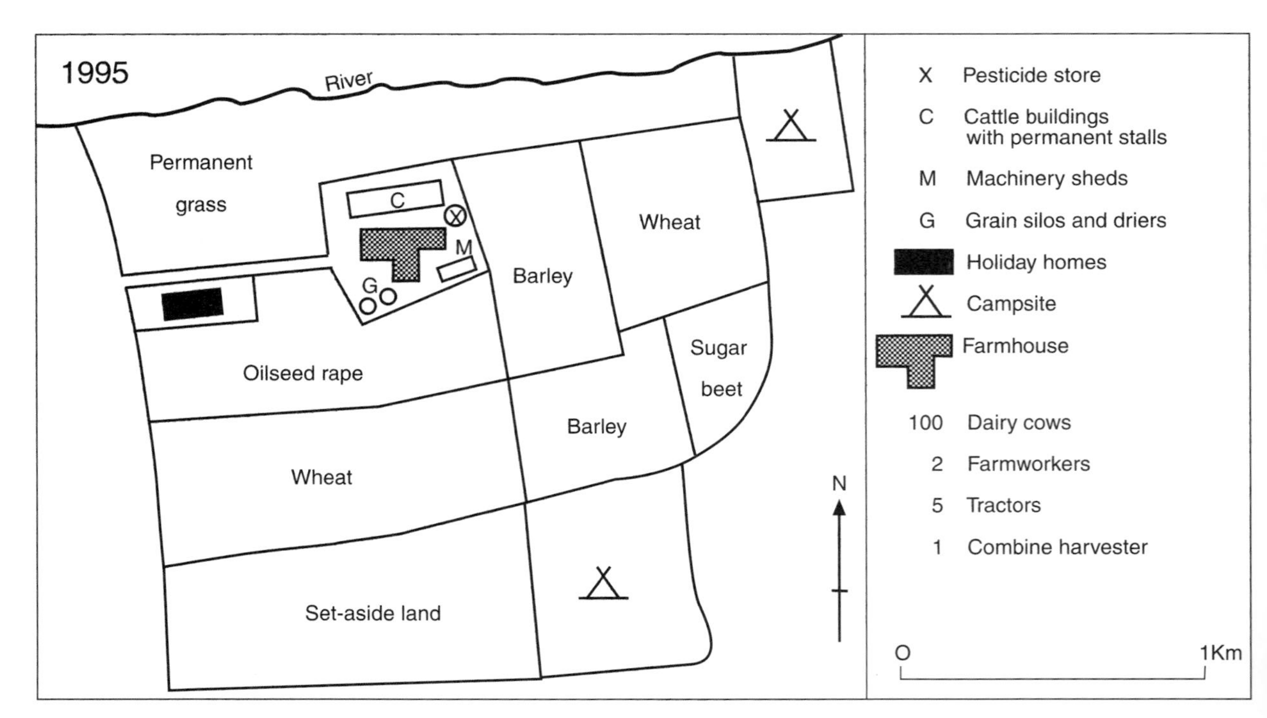


**FIGURE 6.4** Changes in shifting cultivation in the Amazon rain-forest

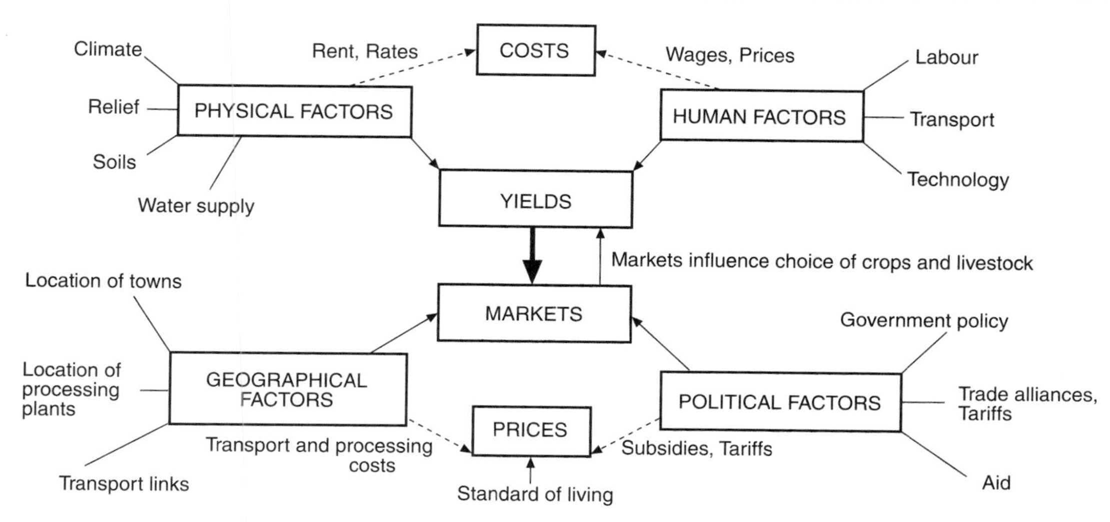


**FIGURE 6.5** Some of the influences on farming

**5** Look at Figure 6.5, which shows some of the influences on farming.

**(a)** Describe how the influences shown in the diagram have affected crop production in an area of extensive commercial farming you have studied. (5)

**(b)** Outline ways in which modern farming methods can result in environmental and landscape damage. (5)

**(10)**

**6** Figure 6.6 shows part of a rice field in an area of intensive subsistence farming.

**(a)** Describe the field and infrastructure patterns in an area of intensive subsistence farming you have studied. (5)

**(b)** Outline ways in which the field and infrastructure patterns of the area you mentioned in (a) can be improved. (5)

**(10)**

**FIGURE 6.6** Part of a rice field in an area of intensive subsistence farming

7 Look at Figure 6.7, which shows a model of the Green Revolution.

**(a)** Describe what is meant by the Green Revolution. (4)

**(b)** Discuss the advantages and disadvantages of the Green Revolution for an area of intensive subsistence farming you have studied. (6)

**(10)**

8 Figure 6.8 shows changes in cereal production and use between 1980 and 1990.

**(a)** Describe the changes shown by the figures. (3)

**(b)** Suggest reasons for the changes shown by the figures. (3)

**(c)** State what action the EU can take to prevent future surpluses. (4)

**(10)**

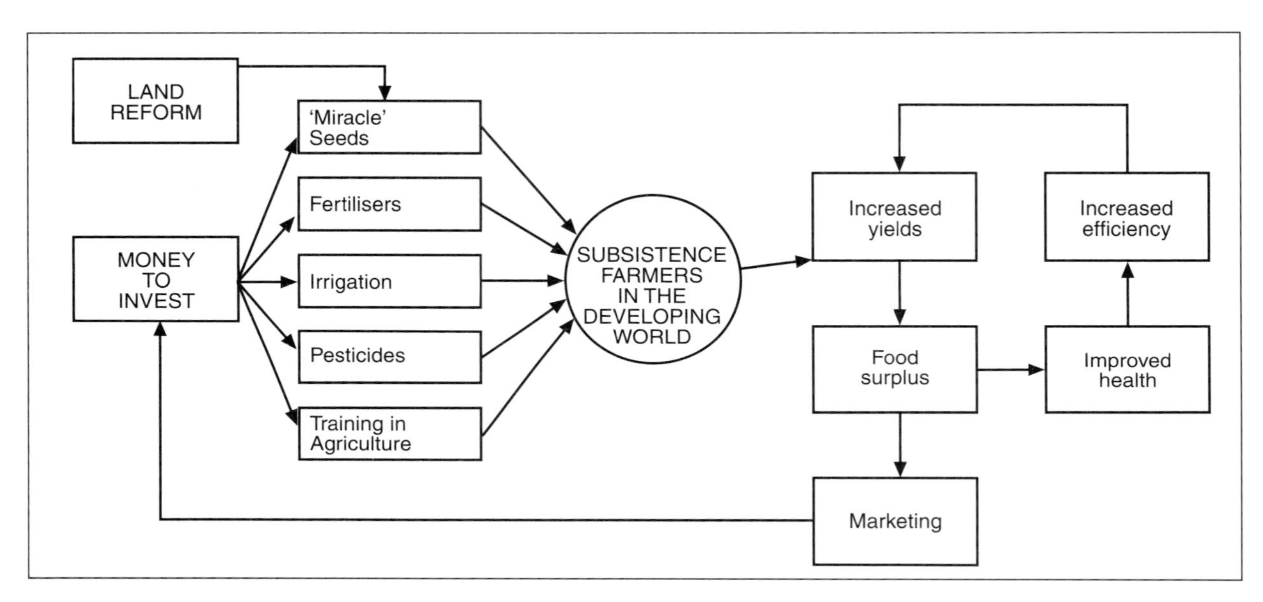


**FIGURE 6.7** A model of the 'Green Revolution'

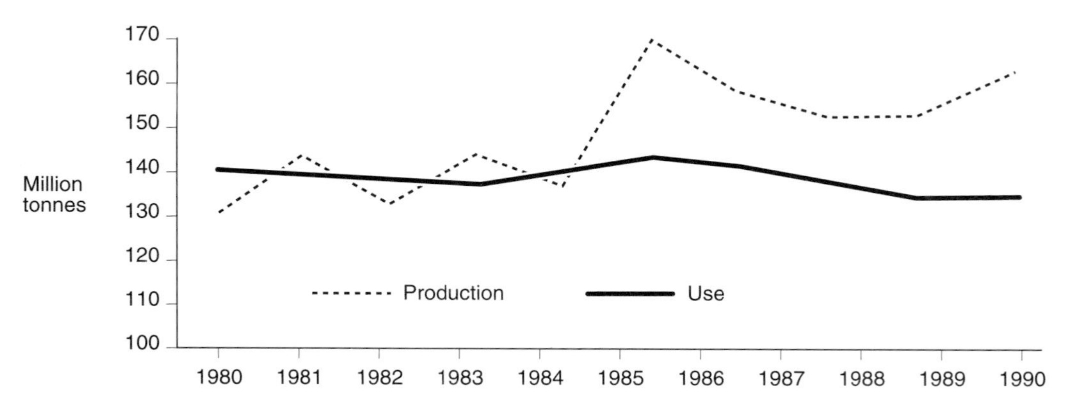


**FIGURE 6.8** Cereal production and use in the European Community between 1980 and 1990

TOTAL NUMBER OF MARKS **80**

# 7
# Industrial Geography

## Key topics

- a case-study of one or more industrial concentrations within the EU;
- the classification of industry – primary, secondary, tertiary and quarternary industries;
- the location of industry – inputs, outputs, linkages, technology and chance factors as well as national and international economic, political and environmental influences;
- changes in industrial location – from water-power to coalfield to coastal to footloose;
- the diversity of industrial landscapes – including a comparison of old and new industrial landscapes;
- the consequences of industrial activity for the environment;
- industrial change – reasons and implications.

## Key skills

- analysing the results of industrial surveys
- using maps to describe and explain industrial locations and landscapes
- annotating and analysing field sketches of old and new industrial landscapes
- analysing the results of employment surveys

## Key words

**agglomeration** a group or concentration of industries

**assembly plant** a factory, such as a car plant, where components are put together

**Assisted Areas** areas which receive help from the government to attract new businesses; this help might include grants, factory buildings and worker training

**brownfield sites** a site which has already been used for industry and therefore has a well-established infrastructure; brownfield sites are often found in inner city areas

**capital** the money required to start or develop an industry

**concealed coalfield** a coalfield where the coal is deep below the ground and requires large modern mines to extract it

**deglomeration** the movement of industries away from an industrial agglomeration or concentration

**de-industrialisation** the decline of industries in a country

**Enterprise Zones** areas where the government tries to attract new industries by offering grants and other forms of assistance

**Euro-core** the industrial and economic centre of the EU, for example: France and Germany

**Euro-periphery** the regions which lie at the edge of the EU, for example: Ireland and Portugal

**exposed coalfield** a coalfield where the coal is at or near the surface and can be obtained from shallow or open-cast mines; reserves of coal in these coalfields have often been worked out

**extractive industry** an industry, such as coalmining, which involves taking resources from the ground

**footloose industry** an industry which has few special site requirements and can be set up in many different locations

**greenfield site** a completely new site for industry, often in a rural area, which offers plenty of space and a dust-free environment

**industrialisation** the growth of industries in a country

**inertia** when an industry remains in a location after the original reasons for it being there have disappeared; for example a steelworks may be built because of local sources of coal and iron-ore but remains on the same site after the coal and iron-ore have been worked out

**infrastructure** the framework of roads, railways, power and water supplies

**industrial estate** a group of factories, often next to a main road or motorway on the edge of a town or city, which offers a clean working environment

**industrial linkages** the ties between industries which require each other's goods and services; for example the different component makers of a car assembly plant

**industrial location** the place where an industry is found

**inputs** the raw materials, energy, buildings, workforce and capital required for industry

**market** a place where goods are sold

**new industrial landscape** new industrial areas, including industrial estates, where factories are kept apart from houses

**old industrial landscape** old industrial areas where factories are mixed in with houses and produce a lot of smoke, dust and noise pollution

**outputs** the finished products which are sent out of a factory

**pollution** smoke, dust, noise and anything else which spoils the environment

**primary industries** industries which involve taking resources from the land or sea; examples include farming, fishing, forestry, quarrying and mining

**quarternary industries** industries which involve providing information, advice and expertise; examples include university and research laboratories

**raw materials** the materials which are used to make other goods; for example iron-ore is an important raw material for making steel

**relocation** moving to a new site

**science park** an industrial estate consisting of science-based industries and research laboratories; science parks are often built next to universities

**secondary industry** industries which involve making or manufacturing finished goods; examples include car-making and shipbuilding

**smokestack industry** a heavy industry, such as steelmaking, which produces a lot of smoke and pollution

**sunrise industry** an expanding industry

**sunset industry** a declining industry

**tertiary industries** industries which involve providing services; examples include banking, retailing and transport services

## TASKS

**1** Figure 7.1 shows some of the EU's most important industrial concentrations.

**(a)** Identify one of these industrial concentrations and describe the types of industrial activities with which it is associated. (6)

**(b)** For the industrial concentration you wrote about in (a), outline the advantages of the region's site and location for the industrial activities which are taking place. (4)

**(10)**

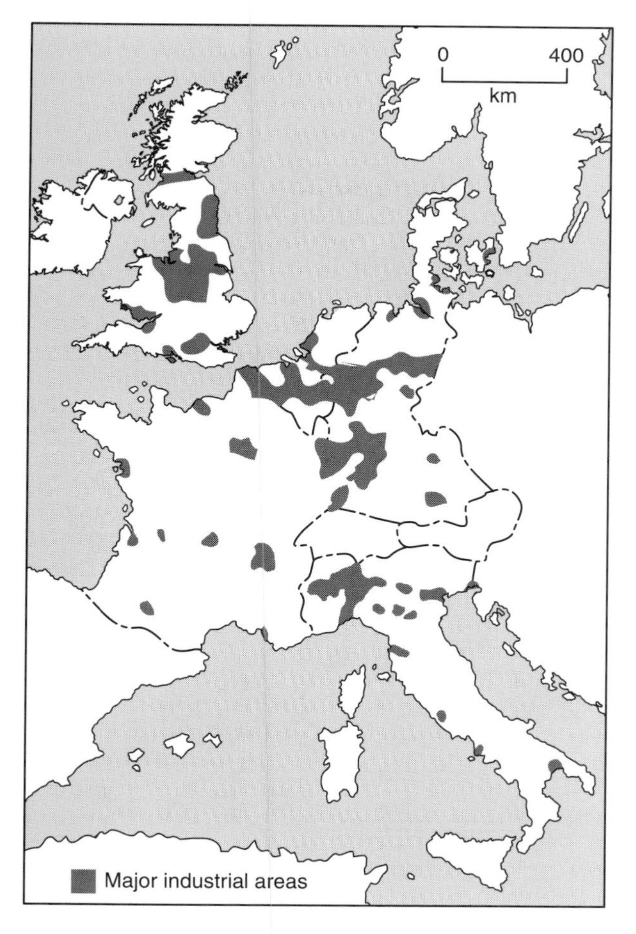


**FIGURE 7.1** Some of the European Community's most important industrial concentrations

**2** Look at Figure 7.2, which shows the location of ironworks and steelworks in South Wales.

**(a)** Describe and explain the location of the original ironworks. (4)

**(b)** For South Wales, or any other steelmaking region in the EU, describe and account for the changes which have affected the location of the region's steelworks. (6)

**(10)**

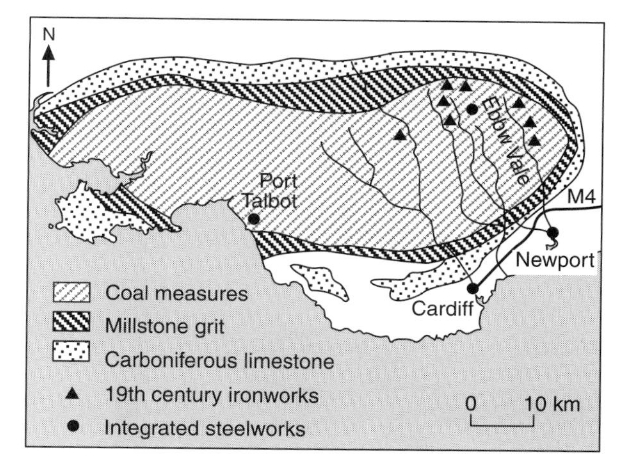


**FIGURE 7.2** The location of iron and steel works in south Wales

**3** Figures 7.3A and 7.3B are two photographs of Scotland's coal mining landscape.

**(a)** For Scotland, or any other region of the EU you have studied, describe how the coal mining industry has changed since the 1960s. (3)

**(b)** Discuss the economic, social and environmental problems which arise when a coal mine closes. (4)

**(c)** Outline the measures which the government can take to assist areas affected by declining industries. (3)

**(10)**

FIGURE 7.3A

FIGURE 7.3B

4 Look at Figure 7.4, which shows an old and a new industrial landscape.

**(a)** Compare and contrast the old and new industrial landscapes in any part of the EU you have studied. (5)

**(b)** Discuss the social and environmental benefits associated with new industrial landscapes. (5)

**(10)**

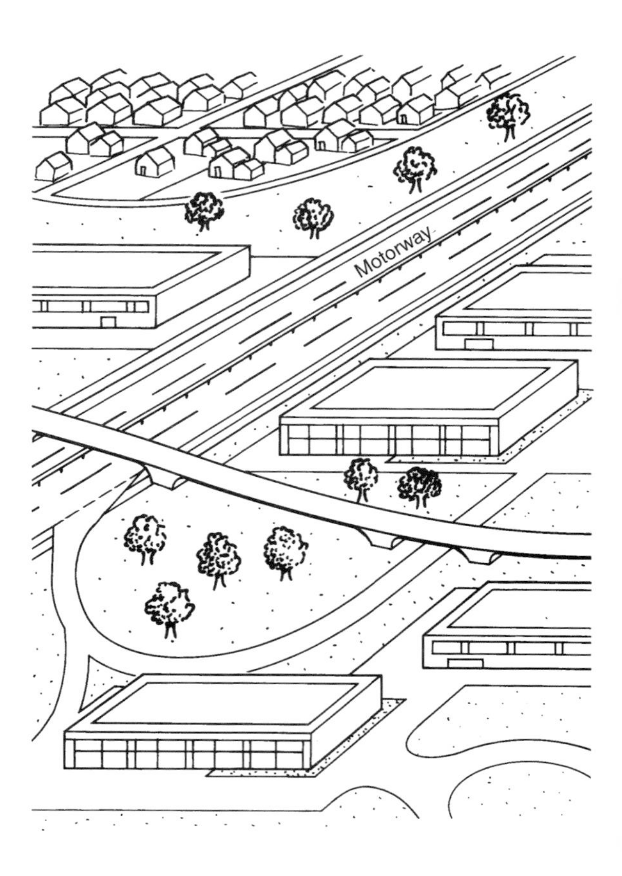


**FIGURE 7.4** An old and new industrial landscape

**5** Figure 7.5 shows the cycle of industrial decline.

**(a)** For any industrial area in the EU which has experienced industrial decline, describe the factors which contributed to the decline. (5)

**(b)** Outline the steps taken to revitalise the area affected by declining industries. (5)

**(10)**

**6** Examine Figure 7.6, which shows the results of an employment survey.

**(a)** State, with the help of examples, the difference between primary, secondary and tertiary industries. (3)

**(b)** Define, with the help of two examples, the meaning of 'quarternary industry'. (3)

**(c)** Account for the differences in the employment patterns of the UK and Portugal. (4)

**(10)**

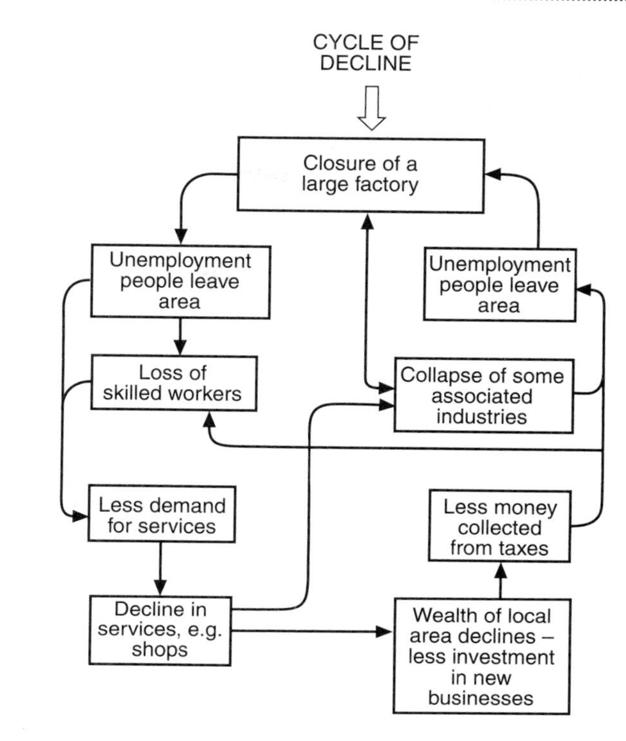


**FIGURE 7.5** The cycle of industrial decline

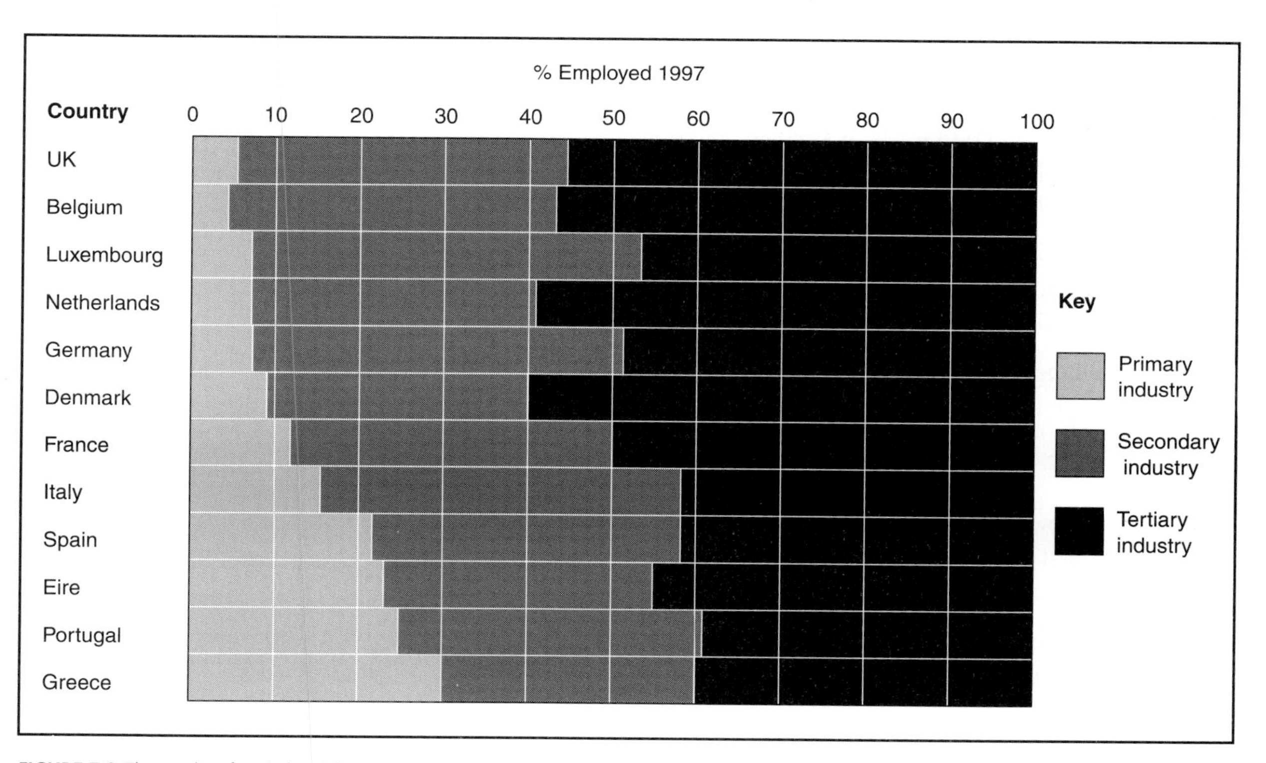


**FIGURE 7.6** The results of an industrial survey

**FIGURE 7.7** The aluminium works at Kinlochleven

**7** Study Figure 7.7, which shows the aluminium works at Kinlochleven in the west Highlands of Scotland, and the OS map of the Kinlochleven area shown on the inside back cover of this book.

**(a)** Draw a labelled sketch to help you describe the site and situation of the Kinlochleven aluminium works. (5)

**(b)** Outline the impact that large industrial works can have on the environment. (5)

**(10)**

| employment sector | % of workforce |
|---|---|
| metal manufacturing | 65 |
| professional | 10 |
| distribution trades | 6 |
| retail | 3 |
| tourism | 6 |
| transport | 2 |
| food and drink | 1 |
| other manufactures | 2 |
| miscellaneous services | 5 |

**FIGURE 7.8** The results of an employment survey in Kinlochleven

**8** Figure 7.8 shows the results of an employment survey taken at Kinlochleven in the west Highlands of Scotland.

**(a)** Present this information in graph form and then describe what the graph shows. (4)

**(b)** In 1996 a decision was taken to close the aluminium plant at Kinlochleven. Discuss the disadvantages, and possible advantages, of this decision for Kinlochleven. (6)

**(10)**

TOTAL NUMBER OF MARKS **80**

# 8
# Urban Geography

## *Key topics*

- case-study of one urban concentration from a developed country, including:
  - its site
  - its situation
  - its functions
  - its urban zones – commercial, industrial and residential landscapes
  - its central business district
  - how and why it grew and changed
  - the redevelopment of its inner city areas
  - its traffic management programmes

## *Key skills*

- analysing urban land use maps (including OS maps)
- analysing urban transects
- analysing survey data (for example commuter flows, pedestrian counts and spheres of influence)
- annotating and analysing field sketches of urban landscapes

## *Key words*

**accessiblity** the ease or difficulty in which a place can be reached; the centres of towns and cities are usually very accessible because they have many routes leading to them

**central business district (cbd)** the centre of a large town or city which is dominated by offices, banks, department stores, entertainment facilities, a town hall and a principal railway station

**commercial zone** an area of banks, building societies and company offices

**concentric zone model** a model of urban structure which is made up of circular zones with the cbd at the centre and residential districts at the edge

**conurbation** a large built-up area consisting of several towns and cities

**decentralisation** the movement of people and businesses away from a central place

**dormitory settlement** a settlement, usually with good transport links, from which people commute to work in nearby towns and cities

**function** the main reason or purpose of a settlement; for example Glasgow is a major route centre and industrial city

**gentrification** the movement of richer people into older, more decayed parts of cities where they improve the buildings and alter the social life

**green belt** a protected area of countryside around a town or city

**grid iron pattern** a street pattern in which the streets run at right angles to each other causing problems for the flow of traffic; grid iron street patterns are often found in older parts of towns and cities

**hamlet** a small settlement with few services

**high order services** services, such as furniture removers, which are only used occasionally

**industrial zone** an area of factories

**inner city** the older residential and industrial areas which can be found on the outskirts of city centres

**inner city decline** the deterioration of houses, shops and factories in the older parts of cities

**linear settlement** a line-shaped settlement which has developed along a road, river or coastline

**low order services** services, such as bus services, which are used frequently

**multiple nuclei model** a model of urban structure which shows urban developments around key points such as railway stations and busy road junctions

**new town** a planned, self-contained settlement which was built to provide new housing and industry for people from overcrowded urban areas

**nucleated settlement** a settlement in which the buildings are grouped together

**pedestrian precinct** a car-free shopping street or area

**redevelopment** changes to improve an area

**residential zone** a housing area

**rural–urban (rurban) fringe** the land at the edge of a town or city which has a mix of rural and urban land uses such as housing and farming

**sector model** a model of urban structure which attempts to explain the urban land use along lines from the city centre

**settlement hierarchy** a list of settlements in order of size

**site** the land on which a settlement has been built

**situation** the position of a settlement in relation to surrounding land and water features

**sphere of influence** the area which is served by a shop, service or settlement

**suburb** a residential area at the edge of a large town or city

**threshold population** the minimum number of people required to support a particular shop or service

**urban renewal** the improvement of run-down inner city areas

## TASKS

1 Examine Figure 8.1, which shows the site and situation of Glasgow.

**(a)** Describe the site and situation of Glasgow, or any other city you have studied from the Developed World. (3)

**(b)** Describe how the site and situation of your chosen city has influenced the city's growth. (3)

**(c)** State the main functions of the city you mentioned in your answers to (a) and (b). Explain how these functions are affected by the city's site and situation. (4)

**(10)**

**FIGURE 8.1** The site and situation of Glasgow

2 Figure 8.2 shows a model of urban growth.

**(a)** Describe the similarities and differences between the model and any city you have studied in the Developed World. (6)

**(b)** Choose two different zones of this model and compare their location and land use. (4)

**(10)**

3 Study Figure 8.3, which shows land use changes in a city.

**(a)** Describe the changes in land use which occur with increasing distance from the city centre. (3)

**(b)** Suggest reasons for these changes. (3)

**(c)** Describe the urban landscape of the central business district. (4)

**(10)**

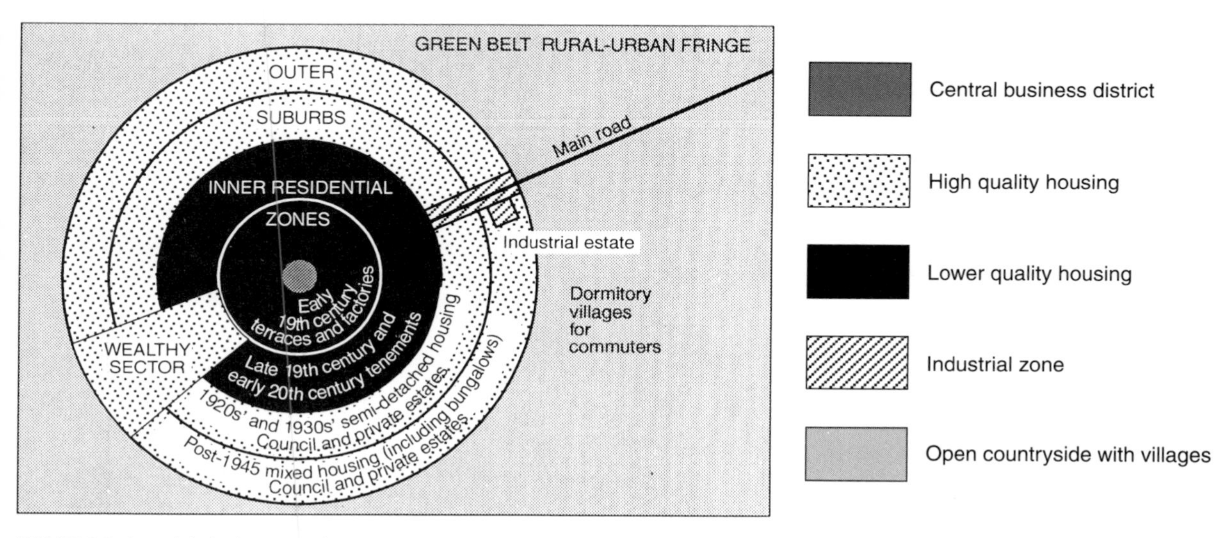


FIGURE 8.2 A model of urban growth

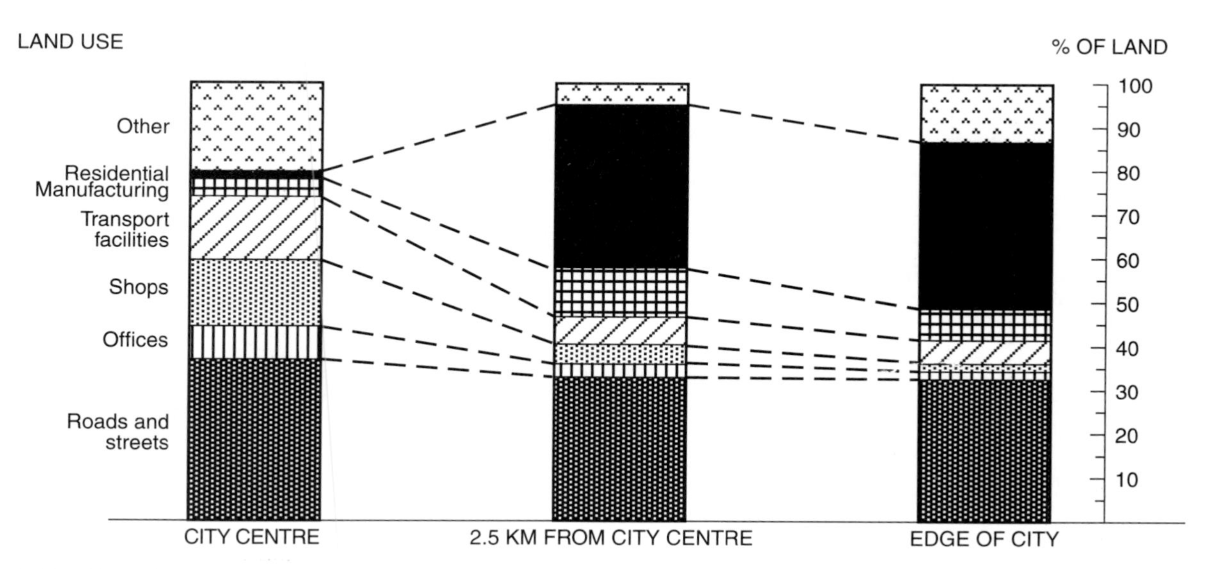


FIGURE 8.3 Land use changes in a city

4 Figures 8.4A, 8.4B, 8.4C and 8.4D show four street patterns.

**(a)** Explain why different street patterns are found in different parts of cities. (4)

**(b)** Compare and contrast the residential environments normally associated with the street patterns shown in Figures 8.4C and 8.4D. (6)

**(10)**

5 Figure 8.5 shows a sketch of an urban landscape. Referring to a city in the Developed World, describe the changes which have taken place in the inner city zone in recent years. Explain why these changes were necessary and comment on the effectivenes of the measures taken. (10)

**(10)**

**FIGURE 8.4A** Street pattern A

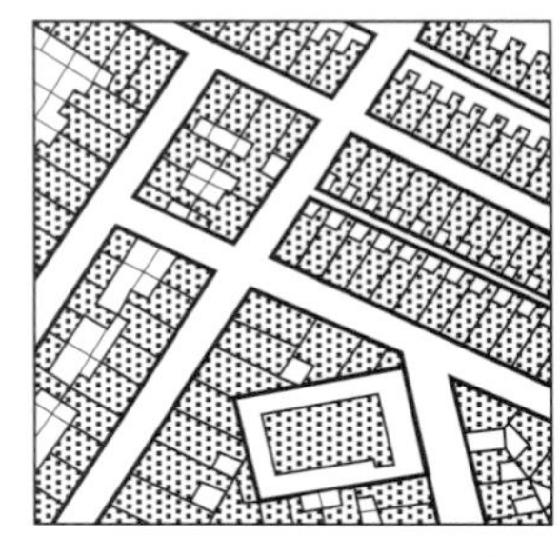

**FIGURE 8.4B** Street pattern B

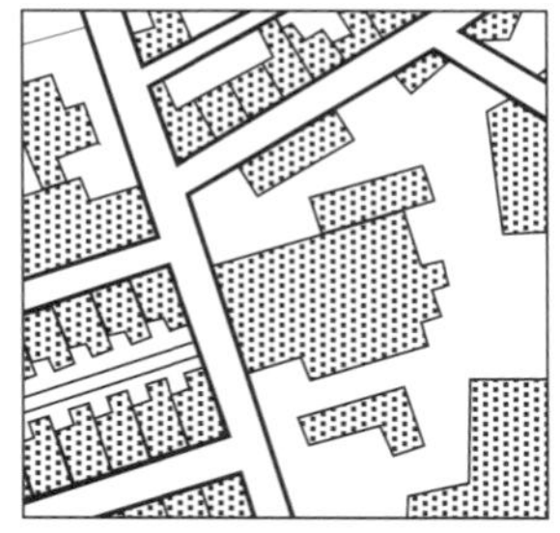

**FIGURE 8.4C** Street pattern C

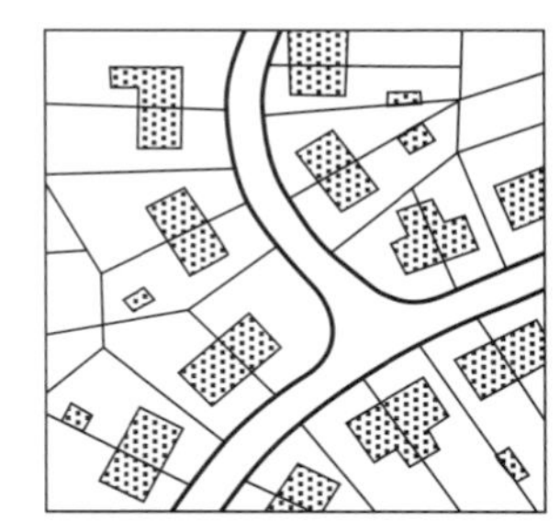

**FIGURE 8.4D** Street pattern D

**FIGURE 8.5** A sketch of an urban landscape

**6** Study Figure 8.6A, which is a photograph of central Stockholm, and Figure 8.6B, an extract from a gazetteer.

**(a)** For Stockholm, or any other city in a developed country, outline the traffic problems which arise because of the city's site, situation and functions. (6)

**(b)** Discuss ways in which these problems might be solved. (4)

**(10)**

Stockholm is spread over 14 islands in the part of south east Sweden where Lake Malaren meets the Baltic Sea. The city is an important port and centre for industries such as engineering and electronics. The Old Town of Stockholm has many historically important and impressive buildings including the Royal Palace and the Parliament Building. Cobbled lanes and winding passages follow the Old Town's original street plan which dates from medieval times. The Old Town is also the centre of Stockholm's thriving tourist industry which attracts millions of visitors and tourists to the city every year.

**FIGURE 8.6B** An extract from a gazetteer

**FIGURE 8.6A** Central Stockholm

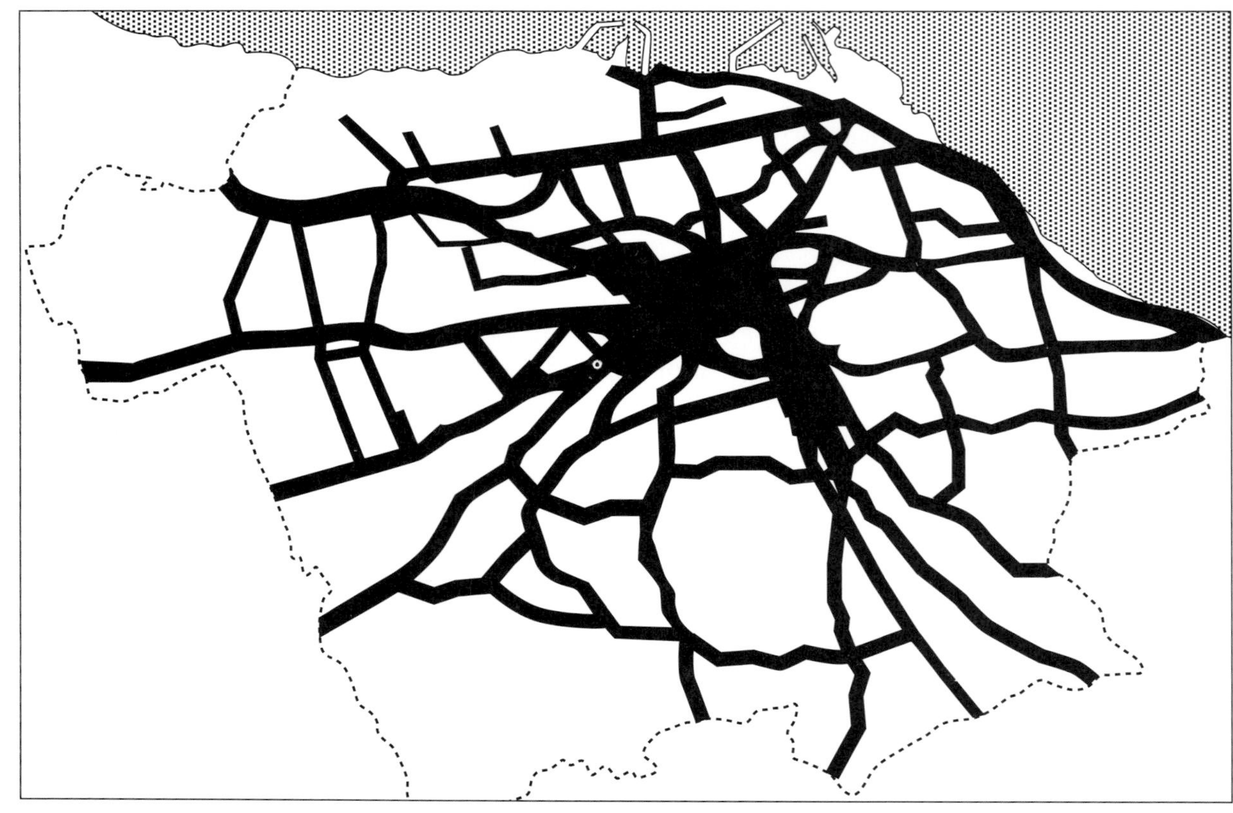

Vehicles per day
0 - 5000
5 - 10000
10 - 15000
15 - 20000
20 - 25000
25 - 30000
over 30000

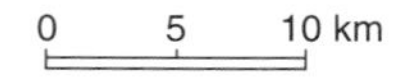


**FIGURE 8.7A** Traffic flows into Edinburgh

7 Study Figure 8.7A, which shows traffic flows into Edinburgh (from a survey taken on a Monday) and Figure 8.7B which shows traffic in the centre of Edinburgh.

**(a)** Describe the flows of traffic into Edinburgh. (3)

**(b)** Suggest reasons for the different traffic flows indicated. (3)

**(c)** Outline the problems which arise when large numbers of cars try to reach the central part of a city. (4)

**(10)**

**FIGURE 8.7B** Traffic in the centre of Edinburgh

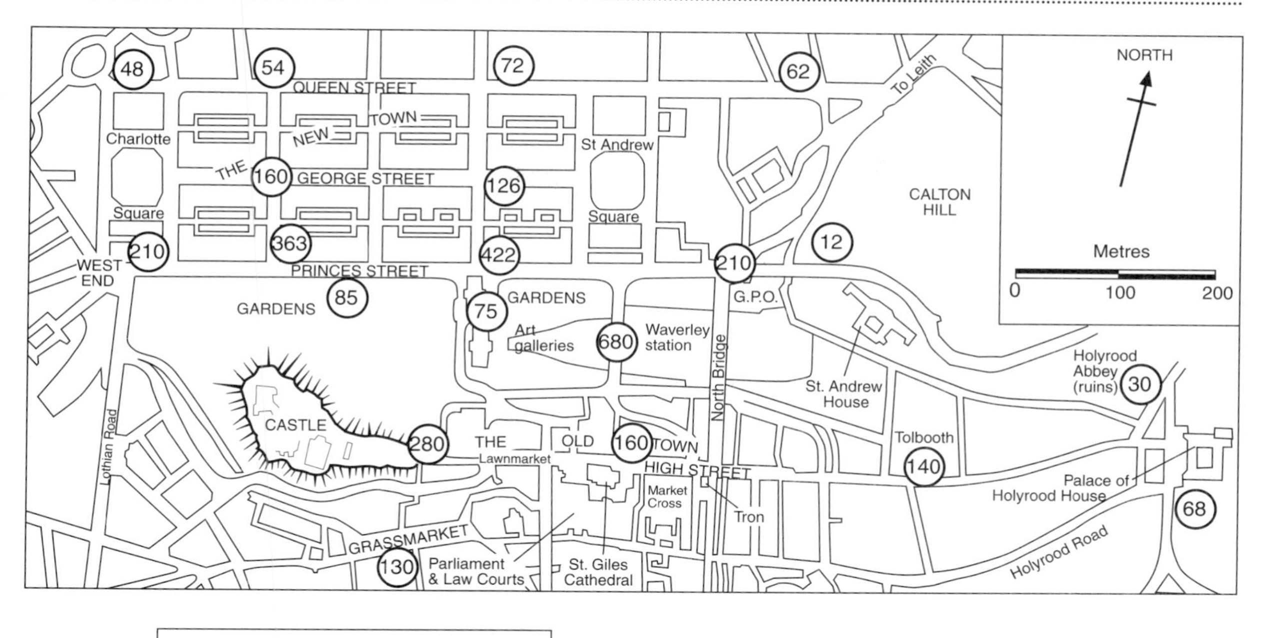


**FIGURE 8.8A** The results of a survey of the distribution of pedestrians in central Edinburgh

8 Study Figures 8.8A, which shows the results of a survey of the distribution of pedestrians in central Edinburgh, and Figure 8.8B which shows pedestrians in the centre of Edinburgh.

**(a)** Discuss the results of the survey. (3)

**(b)** Suggest reasons why some parts of the centre of Edinburgh attract more pedestrians than other parts. (3)

**(c)** What problems arise when large numbers of pedestrians are concentrated in a small area? What measures can be taken to tackle these problems? (4)

**(10)**

TOTAL NUMBER OF MARKS **80**

**FIGURE 8.8B** Pedestrians in the centre of Edinburgh

## 1 ATMOSPHERE

***Question 5***

(a)

Global warming is caused by an increase in gases such as carbon dioxide in the earth's atmosphere. These gases act like glass in a greenhouse – they allow the sun's rays to pass through but trap some of the heat that would otherwise be radiated back into space. This is known as the Greenhouse Effect. Carbon dioxide, which is responsible for more than half of the warming, is the product of a number of different human activities including:

- burning fossil fuels in power stations, factories and motor vehicles;
- destroying trees which soak up the gases when they are growing but release the gases when they are burned down.

Other 'greenhouse' gases include chlorofluorocarbons (CFCs) and methane. CFCs are given off by the coolants used in refrigeration and air cooling systems and by aerosol cans. Methane is a byproduct of a number of agricultural activities including the breakdown of plants in rice fields and marshes.

(6 marks)

(b)

The atmosphere's warming process can be slowed by reducing the number of pollutants entering the atmosphere. This can be achieved by limiting the fossil fuels being burned in power plants, factories and cars and by introducing tougher laws governing the waste gases which can be released into the atmosphere. Carbon dioxide emissions can also be reduced by conserving energy, reducing deforestation and planting more trees. These steps require the co-operation of all countries, as few nations acting on their own can make an impact on global emissions.

(4 marks)

## 2 HYDROSPHERE

***Question 6***

(a)

The photograph shows a series of narrow, steep-sided valleys at the tops of mountains. Water flow in these river courses is rapid because of the steep gradients which are normally found on the upper slopes of mountains. Water levels depend on rainfall – the water level will be low during dry periods and high during wet periods. Waterfalls and potholes are sometimes found in these river courses.

(3 marks)

(b)

The upper courses of rivers are formed by the eroding power of fast-flowing water. Erosion is greatest after heavy rainfalls when larger volumes of water provide the river with more eroding power. Vertical erosion, which makes the valley deeper by eroding material from the river bed, is the dominant process.

Headward erosion sometimes occurs at the beginning of the valley and is a result of rainwash, soil creep and undercutting. Headward erosion results in the valley extending up the slope.

Waterfalls occur where bands of hard, resistant rocks lie across the valleys. The softer rocks are worn away more quickly than the harder rocks, leaving steep drops for the river to flow over. Potholes are created when an uneven river bed results in pebbles being swirled around in the water and gradually producing holes in the bed. This process of erosion is known as corrasion.

The material which is eroded from the valley is transported downstream to other parts of the river channel.

(7 marks)

# 3 LITHOSPHERE

***Question 3***

(a)

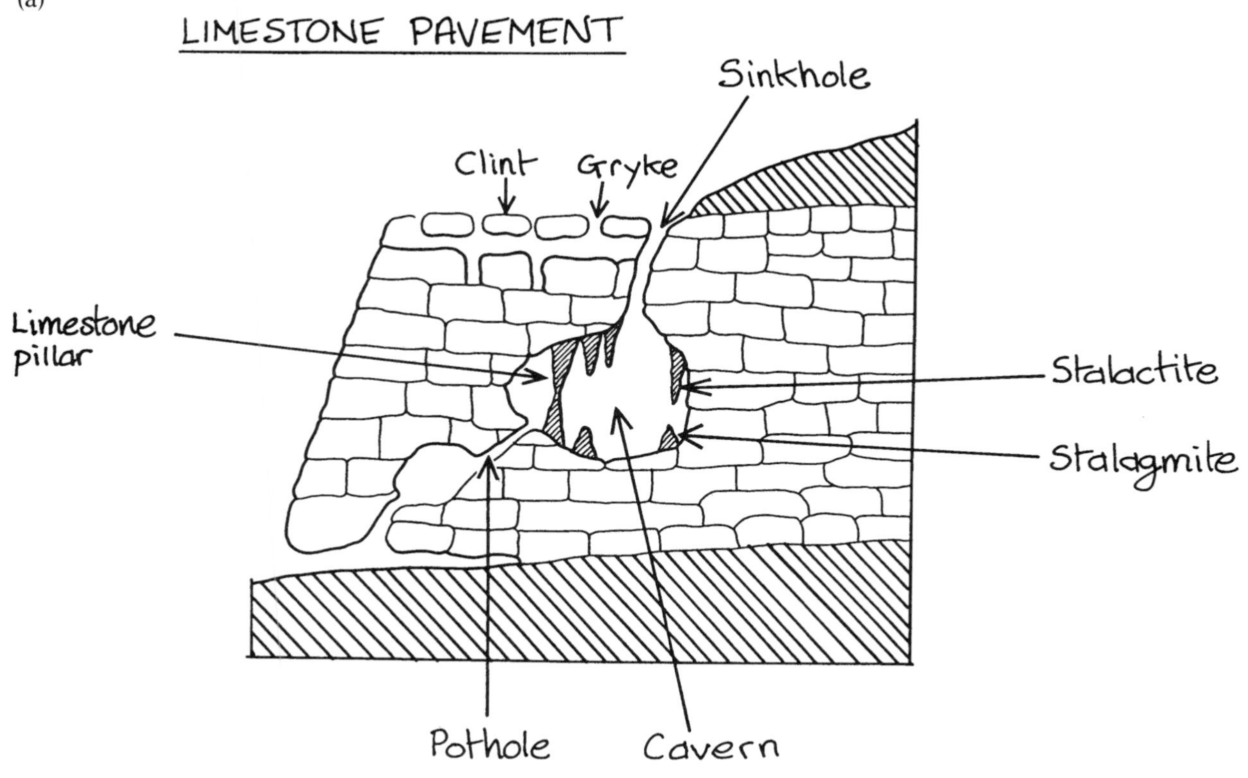


The main surface feature of this carboniferous limestone area is the limestone pavement which is made up of clints and grykes, as shown in the sketch above. The clints are the flat-topped blocks of limestone while the grykes are the spaces or joints separating the clints. Other surface features associated with carboniferous limestone scenery include limestone scars, sinkholes and dry valleys.

The underground features include caverns with stalactites hanging down from the roof, stalagmites projecting upwards from the floor and limestone pillars where stalactites and stalagmites join together to form a continuous column between the floor and roof of the cavern.

(5 marks)

(b)

Limestone is a permeable rock, which means that water can pass through it. The streams flow underground, finding a channel between the limestone's enlarged joints. Although it is a hard rock, limestone dissolves slowly in water. This process, which is known as solution, helps to enlarge the joints in the limestone rocks to produce a series of passages called potholes. Where solution is more active the streams may further enlarge the potholes to create a network of caves.

Limestone pillars are formed when stalactites merge into stalagmites. A stalactite is formed when drops of ground water leave small deposits of calcite hanging from the cave's ceiling. Successive drips cause the stalactite to grow downwards towards the floor of the cave. Water also drips on to the floor of the cave so that deposits of calcite are able to build upwards to produce stalagmites.

(5 marks)

# 4 BIOSPHERE

### *Question 6*

(a)

The types of plants most likely to be found at this site include species which can withstand difficult growing conditions such as poor light, limited moisture and soil which is low in organic matter. At first the site would support pioneer species of grasses, elder and hairy bitter cress as well as plants such as dandelion and willowherb which are able to reproduce themselves very quickly by producing large numbers of seeds. Over time, however, there would be a greater variety of plants including the possible appearance of groundsel and Oxford ragwort. If the site remains derelict for several years then young trees and bushes, including sycamore and rowan, will appear to produce a multi-layered vegetation cover. The later plants sometimes shade out the pioneer and early colonising plants which are unable to compete for the limited light, water and soil resources.

(6 marks)

(b)

Areas of derelict land usually have a wide range of physical conditions, such as variable levels of light, water and humidity. Derelict building walls help to cut out light from many parts of the site and prevent rainwater reaching some areas. Plants in the shaded areas therefore have to be shade-loving species and species which can cope with very dry conditions. Large parts of the site are covered with concrete, with only a few cracks and spaces where plants such as grasses and dandelions can grow. The absence of light, moisture and vegetation over a long period means that soils are often low in organic matter.

(4 marks)

# 5 POPULATION GEOGRAPHY

### *Question 4*

At Stage A both the birth rate and the death rate are high. The birth rate remains quite steady at around 38 births per thousand per year while the death rate fluctuates between 35 and 39 deaths per thousand per year. The population therefore remains stable. The high birth rate is a result of the desire for large families while the high death rate is caused by a combination of factors such as poor diet and poor medical facilities. The upward fluctuations in the death rate are a result of deaths from famines, diseases and wars.

At Stage B the birth rate remains high, at around 38 births per thousand per year, but the death rate falls sharply to around 15 deaths per thousand per year. The population is therefore increasing rapidly. The high birth rate can still be explained by the desire for large families while improvements in health care and food supply account for the declining death rate.

At Stage C the birth rate falls rapidly to around 16 births per thousand per year while the death rate remains low at around 15 deaths per thousand per year. The population therefore grows at a much slower rate. The decline in birth rates can be explained by higher standards of living and the desire for smaller families.

At Stage D the death rate continues to remain low while the birth rate fluctuates slightly between 14 and 17 births per thousand per year. The population therefore remains fairly stationary. The slight fluctuations in the birth rate are a result of a number of different factors including varying income levels which affect peoples' decisions to have babies.

(10 marks)

# 6 RURAL GEOGRAPHY

***Question 1***

(a)

Extensive commercial farming is found in developed countries such as Canada, the USA and the UK. This type of farming requires large areas of low and flat land in temperate parts of the world which are neither too hot nor too cold. Examples include the Prairies of Canada and Fenlands of East Anglia in the United Kingdom.

Intensive peasant farming is found in developing countries such as India, Bangladesh and China in South East Asia. These areas have large areas of flat land with fertile soils and a hot, wet climate. Examples include the fertile valleys of the River Ganges and the River Indus.

Shifting agriculture is restricted to areas of tropical rainforest such as the Amazon rainforest in South America and the Zaire rainforest in central Africa. These areas have a hot, wet climate which allows crops to be grown all year round.

(5 marks)

(b)

Extensive commercial farming involves growing cash crops such as wheat and barley in large, highly-mechanised farms. The use of machines means that few workers have to be employed. The latest farming methods are used including modern chemical fertilisers to provide high crop yields.

Intensive peasant farming involves growing crops such as rice and a few vegetables on small family plots of land. Few machines are used and most of the crops are used to feed the farmers and their families.

Shifting agriculture involves clearing a small patch of rainforest and then planting crops for the farmers and their families to eat. There are no machines and tools are very basic. After a few years the farmers have to move to another part of the rainforest because the original plot has become less fertile.

(5 marks)

# 7 INDUSTRIAL GEOGRAPHY

***Question 3***

(a)

Scotland's coal mining industry has declined dramatically since the 1960s. Hundreds of collieries have closed and over a hundred thousand miners have lost their jobs. Production is now concentrated in a few highly-mechanised collieries and large open-cast pits. The industry is more efficient than it used to be with much higher outputs of coal per miner.

(3 marks)

(b)

The closure of coal mines results in unemployment for miners and other colliery workers. Supply companies and local businesses also suffer because of the loss of custom from the colliery and its workforce. The social problems of mine closures include increased poverty for jobless miners and the movement of workers and their families to jobs in other parts of the country. The closure of a colliery also results in a number of environmental problems including derelict mines and unsightly pit bings.

(4 marks)

(c)

The government can assist areas affected by declining industries by providing grants and loans to attract new businesses to the area and by retraining workers for new jobs in expanding industries. The government can also improve an area's infrastructure, including its transport facilities and power supplies, to help attract new industries.

(3 marks)

# URBAN GEOGRAPHY

***Question 1***

(a)

Glasgow lies on the flood plain of the River Clyde on the west side of Scotland's Midland Valley. The Campsie Fells lie to the north of the city while the Cathkin Braes lie to the south. The city's nucleated site is generally low and flat although the presence of a series of raised beaches and over a hundred drumlins means that some parts of Glasgow have been built on steeper and higher ground. The site's height ranges from around 10 metres to around 110 metres.

(3 marks)

(b)

The large areas of flat and low land in the Glasgow area have enabled the settlement to grow in size and become Scotland's largest city. Urban expansion has been in all directions from its original site on the north side of the River Clyde. Recent growth, however, has been restricted in the north west and the south east because of the Kilpatrick Hills and Cathkin Braes. The general openness of the land has enabled the site to be nucleated in shape rather than linear.

(3 marks)

(c)

Glasgow's situation next to the River Clyde, on the west side of Scotland, enabled the city to develop as a port with important trade links with North America. Glasgow's port function, and the discovery of coal and iron-ore in nearby Lanarkshire, meant that the city also developed as an important industrial city. Although coal mining, steelmaking and many of Glasgow's other older industries have declined, new industries have been attracted to the area. The lack of deep water has meant that the port of Glasgow has been unable to acommodate larger ships and so many of the city's port facilities have closed.

Glasgow's strategic location on the west side of Scotland's Midland Valley means that the city is a major route centre with important motorway, main road, railway and air links with other parts of Scotland, England and overseas.

(4 marks)

# ANSWERING EXAM QUESTIONS

Finally, here are some tips for answering exam questions:

- look at the number of marks for each question – questions with more marks require longer and more detailed answers;
- look for the key instructions in a question – do exactly what the key instructions command: do not, for example, describe when asked to explain; some of the most important key instructions, and their meanings, are listed below;
- include actual examples and case-studies in your answer – especially if the question instructs you to do this;
- keep your answers relevant to the question – you will waste valuable time if you include irrelevant details;
- leave spaces at the end of each answer so that you can write in any additional points you may remember later.

## KEY INSTRUCTIONS IN EXAM QUESTIONS

The following terms often appear in examination questions. It is important to know exactly what each term means and therefore what the question is instructing you to do. Here are some explanations to help you.

**account** give reasons for
**analyse** examine, break down and comment
**annotate** add labels
**assess** weigh up the importance of
**comment** summarise the various points and give an opinion
**compare** outline the similarities and differences
**contrast** highlight the differences
**describe** state the main features or characteristics
**discuss** give different points of view
**examine** look closely
**explain** give reasons
**identify** point out and name
**interpret** bring out the meaning of
**locate** state where a place is
**outline** note the main features
**quote map evidence** give examples, including grid references if possible, from a map
**rank** put in order of size
**select** choose
**suggest** put forward an idea or reason

***How to answer an exam question***

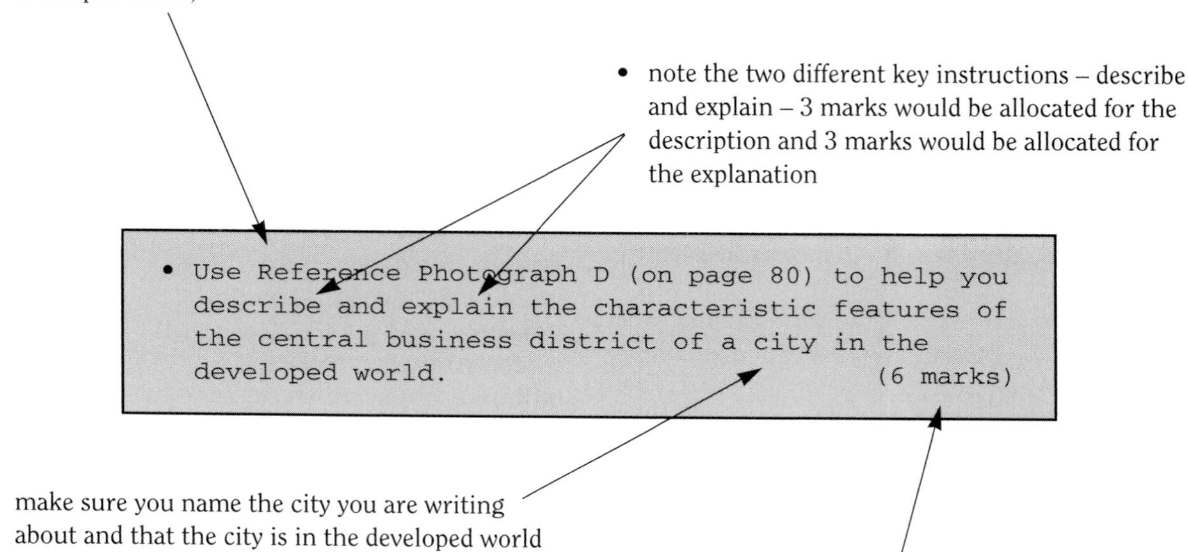


- make sure you name the city you are writing about and that the city is in the developed world

- look at the number of marks allocated for the question – for 6 marks you must write an answer which is at least six sentences long and contains at least six main points

- don't write passages which are irrelevant to the question – you will get no marks for these passages and you will waste valuable time

- leave spaces at the end of each answer so that you can write in any additional points you may remember later

**Reference Photograph A** A rural landscape

**Reference Photograph B** A farming landscape

## SIMULATED FIELDWORK EXERCISES

**1** **(a)** Draw a field sketch of the area shown in Reference Photograph A.

**(b)** Annotate and analyse your completed sketch.

**2** **(a)** Outline one way of collecting land use data for the rural area shown in Reference Photograph B.

**(b)** What problems might you encounter when collecting this data?

**(c)** Suggest one way of presenting rural land use data in a form which makes it easier to provide a detailed land use analysis.

**Reference Photograph C** An industrial landscape

**Reference Photograph D** An urban landscape

**3 (a)** Outline one way of collecting information about the different types of industries and their locations in the area shown in Reference Photograph C.

**(b)** Outline the different ways in which this data could be presented.

**4 (a)** Draw a field sketch of the urban area shown in Reference Photograph D.

**(b)** Annotate and analyse your sketch.